MONOCU[LTURE]
OF THE MIND
*Perspectives on Biodiversity
and Biotechnology*

VANDANA
SHIVA

Zed Books Ltd.
London and New York

TWN
Third World Network
Penang, Malaysia

Monocultures of the Mind
is published by Zed Books Ltd., 7 Cynthia Street,
London, N1 9JF, and Room 400, 175 Fifth Avenue,
New York 10010, USA and by
Third World Network, 228 Macalister Road,
10400 Penang, Malaysia.

First Printing   : 1993
Second Printing : 1995
Third Printing   : 1997
Fourth Printing : 2000

Printed by Jutaprint,
2 Solok Sungai Pinang 3, 11600 Penang, Malaysia.

ISBN 1 85649 217 6 Hb
ISBN 1 85649 218 4 Pb

CONTENTS

INTRODUCTION

THE FIVE ESSAYS in this volume are a selection from my writings over the past decade which reflect on causes of disappearance of diversity and the challenge of conserving it. The main threat to living with diversity comes from the habit of thinking in terms of monocultures; from what I have called 'Monocultures of the Mind'. Monocultures of the mind make diversity disappear from perception, and consequently from the world. The disappearance of diversity is also a disappearance of alternatives – and gives rise to the TINA (there is no alternative) syndrome. How often in contemporary times total uprooting of nature, technology, communities and entire civilisation is justified on the grounds that 'there is no alternative'. Alternatives exist, but are excluded. Their inclusion requires a context of diversity. Shifting to diversity as a mode of thought, a context of action, allows multiple choices to emerge.

The papers are based on participation in movements to protect diversity in nature and culture. My concern about monocultures began with the 'Chipko' movement in the Garhwal Himalaya. The peasant women of Garhwal knew that monoculture pine plantations are not forests, that they cannot perform the multiple functions of providing water and soil conservation services, and providing diverse communities of species for food, fodder, fertiliser, fibre and fuel (the 5-F species in the Chipko language) .

The second experience with the impoverished and impoverishing nature of monocultures was associated with an ecological audit of eucalyptus plantations, especially in the semi-arid zones of Karnataka state where a World Bank social forestry programme was leading to the erosion of farm diversity and a consequent erosion of water and soil, livelihoods and supply of biomass for local use. In 1983, the farmers' movement, the Raitha Sangha, started to uproot eucalyptus seedlings in forest nursery and substitute them with seedlings of diverse species like mango, tamarind, jack fruit, pongamia, etc.

A later study of the Green Revolution in agriculture showed that it was primarily a recipe for the introduction of monocultures and the destruction of diversity. This was also linked to the introduction of centralised control of agriculture and the erosion of decentralised decision-making about cropping patterns. Uniformity and centralisation made for social and ecological vulnerability and breakdown.

Biotechnology and the gene revolution in agriculture and forestry threaten to worsen the trends towards erosion of diversity and centralisation that began with the Green Revolution.

It is in this context of the production of uniformity that the conservation of biodiversity needs to be understood. Conservation of diversity, is, above all, the production of alternatives, of keeping alive alternative forms of production. Protecting native seeds is more than conservation of raw material for the biotechnology industry. The diverse seeds now being pushed to extinction carry within them seeds of other ways of thinking about nature, and other ways of producing for our needs. The critical theme in all the papers is that uniformity and diversity are not just patterns of land use, they are ways of thinking and ways of living. The essays also address the myths that monocultures are essential for solving problems of scarcity and there is no option to destroying diversity to increase productivity. It is not true that, without monoculture tree plantations there will be famines of fuel wood, and without monocultures in agriculture there will be famines of food. Monocultures are in fact a source of scarcity and poverty, both because they destroy diversity and alternatives and also because they destroy decentralised control on production and consumption systems.

Diversity is an alternative to monoculture, homogeneity and uniformity. Living diversity in nature corresponds to a living diversity of cultures. The natural and cultural diversity is a source of wealth and a source of alternatives.

The first essay 'Monocultures of the Mind' was first written for the United Nations University / WIDER programme on 'System of Knowledge as Systems of Power'. It tries to show that monocultures first inhabit the mind, and are then transferred to the ground. Monocultures of the mind generate models of production which destroy diversity and legitimise that destruction as progress, growth and improvement. From the perspective of the monoculture mind, productivity and yields appear to increase when diversity is erased and replaced by uniformity. However, from the perspective of diversity, monocultures are based on a decline in yields and productivity. They are impoverished systems, both qualitatively and quantitatively. They are also highly unstable and non-sustainable systems. Monocultures spread not because they produce more, but because they control more. The expansion of monocultures has more to do with politics and power than with enriching and enhancing systems of biological production. This is as true of the Green Revolution as it is of the gene revolution or the new biotechnologies.

The essays on biodiversity and biotechnology were prepared as Third World Network briefing papers for the UN Conference on Environment and Development, and attempt to show how negotiations on biodiversity cannot be separated from negotiations on biotechnology. They argue that the treatment of biodiversity as mere 'raw material' comes from an anti nature and racist standpoint which treats nature and Third World people's labour as valueless. Biodiversity does not merely get value through biotechnology and genetic engineering performed by 'white men in white lab coats' to quote Pat Mooney. It has intrinsic value and also a high use value for local communities. The paper also sounds a caution against the treatment of biotechnology as an ecological miracle, and a solution to every environmental ill. Biotechnology could be unleashing worse ecological problems than it claims to solve. There is also a deep and unjustified use of power and politics involved in treating biodiversity and its produce as a free and common heritage of mankind when it comes from the Third World, while treating the products of the same biodiversity as patented private property when it is slightly modified in the labs in the North. Diversity as a way of thought would allow a fairer and more just treatment of the contributions of the North and South.

The fourth essay was a paper I wrote and which was published in the book 'Conservation of Biodiversity for Sustainable Development' (edited by O T Sandlund, K Hindar and A H D Brown and published by Scandinavian University Press, Oslo, in 1992). It argues against the distorted notions of obsolescence of living biodiversity inherent to the paradigm of monocultures which goes hand in hand with monopoly rights over the control of biodiversity and threatens us with unanticipated disasters in the form of the gene revolution. The native seed becomes a system of resistance against monocultures and monopoly rights. The shift from uniformity to diversity is essential both ecologically and politically. It is an ecological imperative because only a system based on diversity respects the rights of all species and is sustainable. It is also a political imperative because uniformity goes hand in hand with centralisation, while diversity demands decentered control. Diversity as a way of thought and a way of life is what is needed to go beyond the impoverished monocultures of the mind.

The fifth essay is a critique of The Biodiversity Convention that was finalised in Nairobi in May 1992 and signed by 154 countries during the UNCED 'Earth Summit' in June 1992. This essay points out several flaws that make it likely for the Convention to have negative impacts on the Third World. For the benefit of readers, we have also reproduced the full Convention text as an Appendix.

1 | Monocultures of the Mind

The 'Disappeared' Knowledge Systems

IN ARGENTINA, when the dominant political system faces dissent, it responds by making the dissidents disappear. The 'desparacidos' or the disappeared dissidents share the fate of local knowledge systems throughout the world, which have been conquered through the politics of disappearance, not the politics of debate and dialogue.

The disappearance of local knowledge through its interaction with the dominant western knowledge takes place at many levels, through many steps. First, local knowledge is made to disappear by simply not seeing it, by negating its very existence. This is very easy in the distant gaze of the globalising dominant system. The western systems of knowledge have generally been viewed as universal. However, the dominant system is also a local system, with its social basis in a particular culture, class and gender. It is not universal in an epistemological sense. It is merely the globalised version of a very local and parochial tradition. Emerging from a dominating and colonising culture, modern knowledge systems are themselves colonising.

The knowledge and power nexus is inherent in the dominant system because, as a conceptual framework, it is associated with a set of values based on power which emerged with the rise of commercial capitalism. It generates inequalities and domination by the way such knowledge is generated and structured, the way it is legitimised and alternatives are delegitimised, and by the way in which such knowledge transforms nature and society. Power is also built into the perspective which views the dominant system not as a globalised local tradition, but as a universal tradition, inherently superior to local systems. However, the dominant system is also the product of a particular culture. As Harding observes:

We can now discern the effects of these cultural markings in the discrepancies between the methods of knowing and the interpretations of the world provided by the creators of modern western culture and those characteristics of the rest of us. Western culture's favourite beliefs mirror in sometimes clear and sometimes distorting ways not the world as it is or as we might want it to be, but the social projects of their historically identifiable creators.[1]

The universal/local dichotomy is misplaced when applied to the western and indigenous traditions of knowledge, because the western is a local tradition which has been spread world wide through intellectual colonisation.

The universal would spread in openness. The globalising local spreads by violence and misrepresentation. The first level of violence unleashed on local systems of knowledge is to not see them as knowledge. This invisibility is the first reason why local systems collapse without trial and test when confronted with the knowledge of the dominant west. The distance itself removes local systems from perception. When local knowledge does appear in the field of the globalising vision, it is made to disappear by denying it the status of a systematic knowledge, and assigning it the adjectives 'primitive' and 'unscientific'. Correspondingly, the western system is assumed to be uniquely 'scientific' and universal. The prefix 'scientific' for the modern systems, and 'unscientific' for the traditional knowledge systems has, however, less to do with knowledge and more to do with power. The models of modern science which have encouraged these perceptions were derived less from familiarity with actual scientific practise, and more from familiarity with idealised versions of which gave science a special epistemological status. Positivism, verificationism, falsificationism were all based on the assumption that unlike traditional, local beliefs of the world, which are socially constructed, modern scientific knowledge was thought to be determined without social mediation. Scientists, in accordance with an abstract scientific method, were viewed as putting forward statements corresponding to the realities of a directly observable world. The theoretical concepts in their discourse were in principle seen as reducible to directly verifiable observational claims. New trends in the philosophy and sociology of science challenged the positivist assumptions, but did not challenge the assumed superiority of western systems. Thus, Kuhn, who has shown that science is not nearly as open as is popularly thought, and is the result of the commitment of a specialist community of scientists to presupposed metaphors and paradigms which determine the meaning of constituent terms and concepts, still holds that modern 'paradigmatic' knowledge, is superior to pre-paradigmatic knowledge which represents a kind of primitive state of knowing.[2]

Horton, who has argued against the dominant view of dominant knowledge, still speaks of the 'superior cognitive powers' of the modes of thought of the modern scientific culture which constitute forms of explanation, prediction and control of a power unrivalled in any time and place. This cognitive superiority in his view arises from the 'openness' of modern scientific thinking and the 'closure' of traditional knowledge. As he interprets it, 'In traditional cultures there is no developed awareness of alternatives to the established body of theoretical levels, whereas in the scientifically oriented cultures, such an awareness is highly developed.' [3]

However, the historical experience of non-western culture suggests that it is the western systems of knowledge which are blind to alternatives. The 'scientific' label assigns a kind of sacredness or social immunity to the western system. By elevating itself **above** society and other knowledge systems and by simultaneously excluding other knowledge systems from the domain of reliable and systematic knowledge, the dominant system creates its exclusive monopoly. Paradoxically, it is the knowledge systems which are considered most open, that are, in reality, closed to scrutiny and evaluation. Modern western science is not to be evaluated, it is merely to be accepted. As Sandra Harding has said:

Neither God nor tradition is privileged with the same credibility as scientific rationality in modern cultures.... The project that science's sacredness makes taboo is the examination of science in just the ways any other institution or set of social practises can be examined.[4]

The Cracks of Fragmentation

Over and above rendering local knowledge invisible by declaring it non-existent or illegitimate, the dominant system also makes alternatives disappear by erasing and destroying the reality which they attempt to represent. The fragmented linearity of the dominant knowledge disrupts the integrations between systems. Local knowledge slips through the cracks of fragmentation. It is eclipsed along with the world to which it relates. Dominant scientific knowledge thus breeds a monoculture of the mind by making space for local alternatives disappear, very much like monocultures of introduced plant varieties leading to the displacement and destruction of local diversity. Dominant knowledge also destroys the very **conditions** for alternatives to exist, very much like the introduction of monocultures destroying the very conditions for diverse species to exist.

As metaphor, the monoculture of the mind is best illustrated in the knowledge and practise of forestry and agriculture. 'Scientific' forestry and 'scientific' agriculture, split the plant artificially into

Local Knowledge Systems

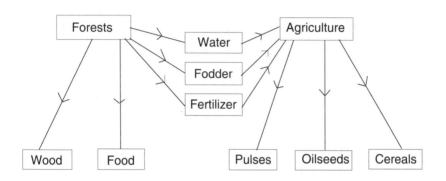

Dominant Knowledge Systems

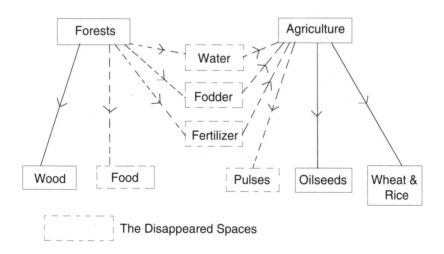

**Fig. 1:   Dominant Knowledge and
The Disappearance of Alternatives.**

separate, non-overlapping domains, on the basis of separate commodity markets to which they supply raw materials and resources. In local knowledge systems, the plant world is not artificially separated between a forest supplying commercial wood and
agricultural land supplying food commodities. The forest and the
field are in ecological continuum, and activities in the forest
contribute to the food needs of the local community, while agriculture itself is modelled on the ecology of the tropical forest. Some
forest dwellers gather food directly from the forest, while many
communities practise agriculture outside the forest, but depend
on the fertility of the forest for the fertility of agricultural land.

In the 'scientific' system which splits forestry from agriculture
and reduces forestry to timber and wood supply, food is no longer
a category related to forestry. The cognitive space that relates
forestry to food production, either directly, or through fertility
links, is therefore erased with the split. Knowledge systems which
have emerged from the food giving capacities of the forest are
therefore eclipsed and finally destroyed, both through neglect and
aggression.

Most local knowledge systems have been based on the life-
support capacities of tropical forests, not on their commercial
timber value. These systems fall in the blind spot of a forestry
perspective that is based exclusively on the commercial exploitation of forests. If some of the local uses can be commercialised, they
are given the status of 'minor products'; with timber and wood
being treated as the 'major products' in forestry. The creation of
fragmented categories thus blinkers out the entire spaces in which
local knowledge exists, knowledge which is far closer to the life of
the forest and more representative of its integrity and diversity.
Dominant forestry science has no place for the knowledge of the
Hanunoo in the Philippines who divide plants into 1,600 categories, of which trained botanists can distinguish only 1,200.[6] The
knowledge base of the cropping systems based on 160 crops of the

Lua tribe in Thailand is not counted as knowledge, either by dominant forestry, which sees only commercial wood, or by dominant agriculture, which sees only chemically intensive agriculture. Food systems based on the forest, either directly, or indirectly are therefore non-existent in the field of vision of a reductionist forestry and a reductionist agriculture even though they have been and still are the sustenance base for many communities of the world. For example, the rainforests of South East Asia supply all the food needs of the Kayan, Kenyah, the Punan Bah, the Penan who gather food from the forest and practise swidden agriculture. The Tiruray people depend on the wild flora of the forests as a major source of food and other necessities.[7] The plant supplies are gathered mostly from the surrounding forest, and some 223 basic plant types are regularly exploited. The most important food items are mushrooms (kulat), ferns (paku) and the hearts of various plants (ubot) which include bamboo shoots, wild palms, and wild bananas. Twenty-five different varieties of fungi are eaten by the Kenyah and 43 varieties are eaten by the Iban.[8] Sago, the staple of the Penan of Borneo, is the starch contained from the pith of a palm tree called the *Eugeissone utilis*. On New Guinea as a whole, (Irian Jaya and Papua New Guinea together) 100,000 sago eaters produce 115,000 metric tons of sago each year.[9] Ethnobotanical work among India's many diverse tribes is also uncovering the deep, systematic knowledge of forests among them. The diversity of forest foods used in India emerges from this knowledge. In South India, a study conducted among the Soliga in the Belirangan hills of Karnataka shows that they use 27 different varieties of leafy vegetables at different times of the year, and a variety of tubers, leaves, fruits and roots are used for their medicinal properties by the tribes. A young illiterate Irula boy from a settlement near Kotagiri identified 37 different varieties of plants, gave their Irula names and their different uses.[10]

In Madya Pradesh, although rice (*Oryza sativa*), and lesser millets (*Panicum miliaceum, Eleusine coracana* and *Paspalum scrobiculatum*) form the staple diet of the tribes, almost all of them supplement it with seeds, grains, roots, rhizomes, leaves and fruits of numerous wild plants which abound in the forests. Grigson noted that famine has never been a problem in Bastar as the tribes have always been able to draw half of their food from the innumerable edible forest products. Tiwari prepared a detailed list of wild plants species eaten by the tribes in Madhya Pradesh. He has listed 165 trees, shrubs and climbers. Of these, the first category contains a list of 31 plants whose seeds are roasted and eaten. There are 19 plants whose roots and tubers are eaten after baking, boiling or processing; there are 17, whose juice is taken fresh or after fermenting; 25, whose leaves are eaten as vegetables, and 10, whose petals are cooked as vegetables. There are 63 plants whose fruits are eaten raw, ripe, roasted or pickled; there are five species of *Ficus* which provide figs for the forest-dwellers. The fruits of the thorny shrub, *Pithcellobium dulce (Inga dulcis)*, also called jungle jalebi, are favourites with the tribes. The sepals of mohwa are greedily eaten and also fermented for liquor. *Morus alba*, the mulberry, provides fruit for both man and birds. Besides, the ber (*Zizyphus mauritania* and *Z oenoplia*) provides delicious fruits, and has been eaten by jungle dwellers from the Mesolithic period onwards.[11]

In non-tribal areas, too, forests provide food and livelihood through critical inputs to agriculture, through soil and water conservation, and through inputs of fodder and organic fertiliser. Indigenous silvicultural practises are based on sustainable and renewable maximisation of all the diverse forms and functions of forests and trees. This common silvicultural knowledge is passed on from generation to generation, through participation in the processes of forest renewal and of drawing sustenance from the forest ecosystems.

In countries like India, the forest has been the source of fertility renewal of agriculture. The forest as a source of fodder and fertiliser has been a significant part of the agricultural ecosystem.

In the Himalaya, the oak forests have been central to sustainability of agriculture. In the western Ghats the 'betta' lands have been central to the sustainability of the ancient spice gardens of pepper, cardamom, and areca nuts. Estimates show that over 50% of the total fodder supply for peasant communities in the Himalaya comes from forest sources, with forest trees supplying 20%.[12] In Dehra Dun, 57% of the annual fodder supply comes from the forests.[13]. Besides fodder inputs, forests also make an important contribution to hill farming in the use of plant biomass as bedding for animals. Forests are the principal source of fallen dry leaf-litter, and lopped green foliage of trees and herbaceous species which are used for animal bedding and composting. Forest biomass, when mixed with animal dung, forms the principal source of soil nutrients for hill agriculture. On one estimate, 2.4 metric tons of litter and manure are used per ha of cultivated land annually.[14] As this input declines, agricultural yields also go down.

The diverse knowledge systems which have evolved with the diverse uses of the forest for food and agriculture were eclipsed with the introduction of 'scientific' forestry, which treated the forest only as a source of industrial and commercial timber. The linkages between forests and agriculture, were broken and the function of the forest as a source of food was no longer perceived.

When the West colonised Asia, it colonised her forests. It brought with it the ideas of nature and culture as derived from the model of the industrial factory. The forest was no longer viewed as having a value itself, in all its diversity. Its value was reduced to the value of commercially exploitable industrial timber. Having depleted their forests at home, European countries started the destruction of Asia's forests. England searched in the colonies for timber for its navy because the oak forest in England were depleted.

The military needs for Indian teak led to the immediate proclamation that wrested the right in the teak trees from the local government and vested it in the East India Company. It was only after more than half a century of uncontrolled destruction of forests by British commercial interests that an attempt was made to control exploitation. In 1865, the first Indian Forest Act (VII of 1865) was passed by the supreme Legislative Council, which authorised the Government to appropriate forests from the local people and manage them as reserved forests.

The introduction of this legislation marks the beginning of what the state and industrial interests have called 'scientific' management. However, for the indigenous people, it amounted to the beginning of the destruction of forests and erosion of peoples' rights to use of the forests. The forests, however, is not merely a timber mine, it is also the source of food for local communities; and with the use of the forests for food and for agriculture, are related diverse knowledge systems about the forest. The separation of forestry from agriculture, and the exclusive focus on wood pro- duction as the objective of forestry led to a creation of one- dimensional forestry paradigm, and the destruction of the multi- dimensional knowledge systems of forest dwellers and forest users.

'Scientific forestry' was the false universalization of a local tradition of forestry which emerged from the narrow commercial interests which viewed the forest only in terms of commercially valuable wood. It first reduced the value of diversity of life in the forest to the value of a few commercially valuable species, and further reduced the value of these species to the value of their dead product – wood. The reductionism of the scientific forestry para- digm created by commercial industrial interests violates both the integrity of the forests and the integrity of forest cultures who need the forests in its diversity to satisfy their needs for food, fibre and shelter.

The existing principles of scientific forest management leads to the destruction of the tropical forest ecosystem because it is based on the objective of modelling the diversity of the living forest on the uniformity of the assembly line. Instead of society being modelled on the forest as is the case for forest cultures, the forest is modelled on the factory. The system of 'scientific management', as has been practised over a century is thus a system of tropical deforestation, which transforms the forest from a renewable to a non-renewable resource. Tropical timber exploitation thus becomes like mining, and tropical forests become a timber mine. According to a FAO estimate, at current rates of exploitation, the forests of tropical Asia will be totally exhausted by the turn of the century.

The tropical forests, when modelled on the factory and used as a timber mine, becomes a non-renewable resource. Tropical peoples also become a dispensable and historical waste. In place of cultural and biological pluralism, the factory produces non-sustainable monocultures in nature and society. There is no place for the small, no value for the insignificant. Organic diversity gives way to fragmented atomism and uniformity. The diversity must be weeded out, and the uniform monocultures – of plants and people – must now be externally managed because they are no longer self-regulated and self-governed. Those that do not fit into the uniformity must be declared unfit. Symbiosis must give way to competition, domination and dispensability. There is no survival possible for the forest or its people when they become feedstock for industry. The survival of the tropical forests depends on the survival of human societies modelled on the principles of the forest. These lessons for survival do not come from text of 'scientific forestry'. They lie hidden in the lives and beliefs of the forest peoples of the world.

There are in Asia today two paradigms of forestry – one life-enhancing, the other life-destroying. The life-enhancing para-

digm emerges from the forest and the forest communities – the
life-destroying from the market. The life-enhancing paradigm
creates a sustainable, renewable forest system, supporting and
renewing food and water systems. **The maintenance of conditions for renewability is the primary management objective of
the former.** The maximising of profits through commercial extraction is the primary management objective of the latter. Since
maximizing profits is consequent upon destruction of conditions
of renewability, the two paradigms are cognitively and ecologically incommensurate. Today, in the forests of Asia the two
paradigms are struggling against each other. This struggle is very
clear in the two slogans on the utility of the Himalayan forests, one
emanating from the ecological concepts of Garhwali women, the
other from the sectoral concepts of those associated with trade in
forest products. When Chipko became an ecological movement in
1977 in Adwani, the spirit of local science was captured in the
slogan:

What do the forests bear?
Soil, water and pure air.

This was the response to the commonly accepted slogan of the
dominant science:

What do the forests bear?
Profit on resin and timber.

The insight in these slogans represented a cognitive shift in the
evolution of Chipko. The movement was transformed qualitatively from being based merely on conflicts over resources in
involving conflicts over scientific perceptions and philosophical
approaches to nature. This transformation also created that element
of scientific knowledge which has allowed Chipko to reproduce
itself in different ecological and cultural contexts. The slogan has
become the scientific and philosophical message of the movement,

and has laid the foundations of an alternative forestry science, oriented to the public interest and ecological in nature. The commercial interest has the primary objective of maximising exchange value through the extraction of commercially valuable species. Forest ecosystems are therefore reduced to the timber of commercially valuable species.

'Scientific forestry' in its present form is a reductionist system of knowledge which ignores the complex relationships within the forest community and between plant life and other resources like soil and water. Its pattern of resource utilisation is based on increasing 'productivity' on these reductionist foundations. By ignoring the system's linkages within the forest ecosystem, this pattern of resource use generates instabilities in the ecosystem and leads to a counterproductive use of natural resources at the ecosystem level. The destruction of the forest ecosystem and the multiple functions of forest resources in turn hurts the economic interests of those sections of society which depend on the diverse resource functions of the forests for their survival. These include soil and water stabilisation and the provision of food, fodder, fuel, fertiliser, etc.

Forest movements like Chipko are simultaneously a critique of reductionist 'scientific' forestry and an articulation of a framework for an alternative forestry science which is ecological and can safeguard the public interest. In this alternative forestry science, forest resources are not viewed as isolated from other resources of the ecosystem. Nor is the economic value of a forest reduced to the commercial value of timber.

'Productivity', 'yield', and 'economic value' are defined for the integrated ecosystem and for multi-purpose utilisation. Their meaning and measure is therefore entirely different from the meaning and measure employed in reductionist forestry. Just as in the shift from Newtonian to Einsteinian physics, the meaning of 'mass' changed from a velocity-independent to a velocity-dependent term, in a shift from reductionist forestry to ecological forestry, all scientific terms are changed from ecosystem-independent to ecosystem-dependent ones. Thus, while for tribes and other forest communities a complex ecosystem is productive in terms of herbs, tubers, fibre and genepool, etc., for the forester, these components of the forests ecosystem are useless, unproductive, dispensable.

The Chipko and Appiko Movements are movements of agricultural communities against the destruction of the forests that support agriculture. The timber blockades of the Penan and other tribes of Sarawak are struggles of forest peoples against systems of forest management which destroy the forest and its people. According to the tribes:

This is the land of our forefathers, and their forefathers before them. If we don't do something now to protect the little that is left, there will be nothing for our children. Our forests are mowed down, the hills are levelled, the sacred graves of our ancestors have been desecrated, our waters and our streams are contaminated, our plant life is destroyed, and the forest animals are killed or have run away. What else can we do now but to make our protests heard, so that something can be done to help us?

AVEK MATAI AME MANEU MAPAT (until we die we will block this road).[15]

The Destruction of Diversity as 'Weeds'

The destruction of biological diversity is intrinsic to the very manner in which the reductionist forestry paradigm conceives of the forest. The forest is defined as 'normal' according to the objective of managing the forest for maximising production of marketable timber. Since the natural tropical forest is characterised by richness in diversity, including the diversity of non-marketable, non-industrial species, the 'scientific forestry' paradigm declares the natural forest as 'abnormal'. In Sclich's words, forest management implies that 'the abnormal conditions are to be removed'[16] and according to Troup,

> *The attainment of the normal forest from the abnormal condition of our existing natural forest, involves a certain temporary sacrifice. Generally speaking, the more rapid the change to the normal state, the greater the sacrifice, for example, the normal forests can be attained in one rotation by a series of clear fellings with artificial regeneration, but in an irregular, uneven-aged forest this means the sacrifice of much young growth which may be unsaleable. The question of minimising the sacrifice involved in introducing order out of chaos is likely to exercise our minds considerably in connection with forest management.*[17]

The natural forest, in its diversity, is thus seen as 'chaos'. The man-made forest is 'order'. 'Scientific' management of forests therefore has a clear anti-nature bias, and a bias for industrial and commercial objectives, for which the natural forest must be sacrificed. Diversity thus gives way to uniformity of even-aged, single species stands, and this uniformity is the ideal of the normal forestry towards which all silvicultural systems aim. The destruction and dispensability of diversity is intrinsic to forest management guided by the objective of maximising commercial wood production, which sees non-commercial parts and relationships of a forest ecosystem as valueless – as weeds to be destroyed. Nature's wealth characterised by diversity is destroyed to create commercial wealth characterised by uniformity.

In biological terms, tropical forests are the most productive biological systems on our planet. A large biomass is generally characteristic of tropical forests. The quantities of wood especially are large in tropical forests and average about 300 tons per ha. Compared with about 150 tons per ha. for temperate forests.

However, in the reductionist commercial forestry, the overall productivity is not important, nor are the functions of tropical forests in the survival of tropical peoples. It looks only for the industrially useful species that can be profitably marketed and measures productivity in terms of industrial and commercial biomass alone. It sees the rest as waste and weeds. As Bethel, an international forestry consultant states, referring to the large biomass typical of the forests of the humid tropics:

*It must be said that from a standpoint of industrial material supply, this is relatively unimportant. The important question is how much of this biomass represents trees and parts of trees of **preferred species that can be profitably marketed**...... By today's utilisation standards, **most of the trees, in these humid tropical forests are, from an industrial materials standpoint, clearly weeds.***[18]

The industrial materials standpoint is the capitalist reductionist forestry which splits the living diversity and democracy of the forest into commercially valuable dead wood and destroys the rest as 'weeds' and 'waste'. This 'waste' however is the wealth of biomass that maintains nature's water and nutrient cycles and satisfies needs of food, fuel, fodder, fertiliser, fibre and medicine of agricultural communities.

Just as 'scientific' forestry excludes the food producing functions of the forest, and destroys the forest diversity as 'weeds', 'scientific' agriculture too destroys species which are useful as food, even though they may not be useful on the market.

The Green Revolution has displaced not just seed varieties but entire crops in the Third World. Just as people's seeds were declared 'primitive' and 'inferior' by the Green Revolution ideology, food crops were declared 'marginal', 'inferior' and 'coarse grained'. Only a biased agricultural science rooted in capitalist patriarchy could declare nutritious crops like ragi and jowar as inferior. Peasant women know the nutrition needs of their families and the nutritive content of the crops they grow. Among foodcrops they prefer those with maximum nutrition to those with a value in the market. What have usually been called 'marginal crops' or 'coarse grains' are nature's most productive crops in terms of nutrition. That is why women in Garhwal continue to cultivate mandua and women in Karnataka cultivate ragi inspite of all attempts by state policy to shift to cash crops and commercial foodgrains, to which all financial incentives of agricultural 'development' are tied. Table 1 illustrates how what the Green Revolution has declared 'inferior' grains are actually superior in nutritive content to the so-called 'superior' grains, rice and wheat. A woman in a Himalaya village once told me, 'Without our mandua and jhangora, we could not labour as we do. These grains are our source of health and strength'.

Not being commercially useful, people's crops are treated as 'weeds' and destroyed with poisons. The most extreme example of this destruction is that of bathua, an important green leafy vegetable, with a very high nutritive value and rich in Vitamin A, which

Table 1				
Nutritional content of different foodcrops				
	Protein (gms)	Minerals (100 gms)	Ca (mg)	Fe (100 gms)
Bajra	11.6	2.3	42	5.0
Ragi	7.3	2.7	344	6.4
Jowar	10.4	1.6	25	5.8
Wheat (milled)	11.8	0.6	23	2.5
Rice (milled)	6.8	0.6	10	3.1

grows as an associate of wheat. However, with intensive chemical fertilizer use bathua becomes a major competitor of wheat and has been declared a `weed' that is killed with herbicides. Forty thousand children in India go blind each year for lack of Vitamin A, and herbicides contribute to this tragedy by destroying the freely available sources of vitamin A. Thousands of rural women who make their living by basket and mat-making, with wild reeds and grasses, are also losing their livelihoods because the increased use of herbicide is killing the reeds and grasses. The introduction of herbicide-resistant crops will increase herbicide use and thus increase the damage to economically and ecologically useful plant species. Herbicide resistance also excludes the possibility of rotational and mixed-cropping, which are essential for a sustainable and ecologically balanced agriculture, since the other crops would be destroyed by the herbicide. US estimates now show a loss of US $4 billion per annum due to loss as a result of herbicide spraying. The destruction in India will be far greater because of higher plant diversity, and the prevalence of diverse occupations based on plants and biomass.

Strategies for genetic engineering resistance which are destroying useful species of plants can also end up creating superweeds. There is an intimate relationship between weeds and crops, especially in the tropics where weedy and cultivated varieties have genetically interacted over centuries and hybridize freely to produce new varieties. Genes for herbicides tolerance, that genetic engineers are striving to introduce into crop plants may be transferred to neighbouring weeds as a result of naturally, occurring gene transfer.

Scarcities of locally useful plant varieties have been created because the dominant knowledge systems discounts the value of local knowledge and declares locally useful plants to be 'weeds'. Since dominant knowledge is created from the perspective of increasing commercial output, and responds only to values on the market, it cannot see the values assigned to plant diversity by local perceptions. Diversity is thus destroyed in plant communities and forest and peasant communities, because in commercial logic it is not 'useful'. And as Cotton Mather, the famous witch hunter of Salem, Massachusetts had stated, 'what is not useful is vicious.' It must therefore be destroyed. When what is useful and what is not is determined one-sidedly, all other systems of determining value are displaced.

Declaring a locally useful species a weed is another aspect of the politics of disappearance by which the space of local knowledge shrinks out of existence. The one-dimensional field of vision of the dominant system perceives only one value, based on the market, and it generates forestry and agricultural practises which aim at maximising that value. Related to the destruction of diversity as valueless is the inevitability of the monoculture as the only 'productive' and 'high yield' system.

'Miracle Trees' and 'Miracle Seeds'

The one-dimensional perspective of dominant knowledge is rooted in the intimate links of modern science with the market. As multidimensional integrations between agriculture and forestry at the local level are broken, new integrations between non local markets and local resources are established. Since economic power is concentrated in these remote centres of exploitation, knowledge develops according to the linear logic of maximising flow at the local level. The integrated forest and farm gives way to the separate spheres of forestry and agriculture. The diverse forest and agricultural ecosystems are reduced to 'preferred' species by selective annihilation of species diversity which is not 'useful' from the market perspective. Finally, the 'preferred' species themselves have to be engineered and introduced on the basis of 'preferred' traits. The natural, native diversity is displaced by introduced monocultures of trees and crops.

In forestry, as the paper and pulp industry rose in prominence, pulp species became the 'preferred' species by the dominant knowledge system. Natural forests were clear-felled and replaced by monocultures of the exotic **Eucalyptus** species which were good for pulping. However, 'scientific' forestry did not project its practise as a particular response to the particular interest of the pulp industry. It projected its choice as based on a universal and objective criteria of 'fast growth' and 'high yields'. In the 1980s, when the concern about deforestation and its impact on local communities and ecological stability created the imperative for afforestation programmes, the eucalyptus was proposed world-wide as a 'miracle' tree. However, local communities everywhere seemed to think otherwise.

The main thrust of conservation struggles like Chipko is that forests and trees are life-support systems, and should be protected and regenerated for their biospheric functions. The monoculture mind on the other hand sees the natural forest and trees as 'weeds' and converts even afforestation into deforestation and desertification. From life-support systems, trees are converted into green gold – all planting is motivated by the slogan, 'Money grows on trees'. Whether it is schemes like social forestry or wasteland development, afforestation programmes are conceived at the international level by 'experts' whose philosophy of tree-planting falls within the reductionist paradigm of producing wood for the market, not biomass for maintaining ecological cycles or satisfying local needs of food, fodder and fertilizer. All official programmes of afforestation, based on heavy funding and centralised decision making, act in two ways against the local knowledge systems – they destroy the forest as a diverse and self-producing system, and destroy it as commons, shared by a diver-sity of social groups with even the smallest having rights, access and entitlements.

'Social' forestry and the 'miracle' tree

Social forestry projects are a good example of single-species, single commodity production plantations, based on reductionist models which divorce forestry from agriculture and water management, and seeds from markets.

A case study of World Bank-sponsored social forestry in Kolar district of Karnataka[19] is an illustration of reductionism and maldevelopment in forestry being extended to farmland. Decentred agro-forestry, based on multiple species and private and common tree-stands, has been India's age-old strategy for maintaining farm productivity in arid and semi arid zones. The honge, tamarind, jackfruit and mango, the jola, gobli, kagli and bamboo traditionally provided food and fodder, fertilizer and pesticide, fuel and small timber. The backyard of each rural home was a nursery, and each peasant a silviculturalist. The invisible, decentred agro-forestry model was significant because the humblest of species and the smallest of people could participate in it, and with space for the small, everyone was involved in protecting and planting.

The reductionist mind took over tree-planting with 'social forestry'. Plans were made in national and international capitals by people who could not know the purpose of the honge and the neem, and saw them as weeds. The experts decided that indigenous knowledge was worthless and 'unscientific', and proceeded to destroy the diversity of indigenous species by replacing them with row after row of eucalyptus seedlings in polythene bags, in government nurseries. Nature's locally available seeds were laid waste; people's locally available knowledge and energies were laid waste. With imported seeds and expertise came the import of loans and debt and the export of wood, soils and people. Trees, as a living resource, maintaining the life of the soil and water and of local people, were replaced by trees whose dead wood went straight to a pulp factory hundreds of miles away. The smallest farm became a supplier of raw material to industry and ceased to be a supplier of food to local people. Local work, linking the trees to the crops, disappeared and was replaced by the work of brokers and middlemen who brought the eucalyptus trees on behalf of industry. Industrialists, foresters and bureaucrats loved the euca-lyptus because it grows straight and is excellent pulp wood, unlike the honge which shelters the soil with its profuse branches and dense canopy and whose real worth is as a living tree on a farm.

The honge could be nature's idea of the perfect tree for arid Karnataka. It has rapid growth of precisely those parts of the tree, the leaves and small branches, which go back to the earth, enrich-ing and protecting it, conserving its moisture and fertility. The eucalyptus, on the other hand, when perceived ecologically, is unproductive, even negative, because this perception assesses the 'growth' and 'productivity' of trees in relation to the water cycle and its conservation, in relation to soil fertility and in relation to human needs for food and food production. The eucalyptus has destroyed the water cycle in arid regions due to its high water demand and its failure to produce humus, which is nature's mechanism for conserving water.

Most indigenous species have a much higher biological pro-ductivity than the eucalyptus, when one considers water yields and water conservation. The non-woody biomass of trees has never been assessed by forest measurements and quantification within the reductionist paradigm, yet it is this very biomass that functions in conserving water and building soils. It is little wonder that Garhwal women call a tree 'dali' or branch, because they see the productivity of the tree in terms of its non-woody biomass which functions critically in hydrological and nutrient cycles within the forest, and through green fertilizer and fodder in cropland.

Eucalyptus

The most powerful argument in favour of the expansion of Eucalyptus is that it is faster growing than all indigenous alternatives. This is quite clearly untrue for ecozones where Eucalyptus has had no productivity due to pest damage. It is also not true for zones with poor soils and poor water endowment, as the reports on yields make evident. Even where biotic and climatic factors are conducive to good growth, Eucalyptus cannot compete with a number of indigenous fast growing species. When tall scientific claims about the growth rate of Eucalyptus were being used to convert rich natural forests to Eucalyptus monoculture plantations, on the ground of the improvement of the productivity of the site, the Central Silviculturist and Director of Forestry Research of the Forest Research Institute (FRI) had categorically stated that 'some indigenous species are as fast growing as, and in some cases even more than, the much coveted Eucalyptus.' In justification he provided a long list of indigenous fast growing species which had growth rates exceeding that of the Eucalyptus which under the best conditions, is about 10 CuM/ha/yr and on average is about 5 CuM/ha/yr (Table 2). Indigenous trees are those trees which are native to the Indian soil or are exotics that have been naturalised over thousands of years.

This data based on forest plantations does not include fast-growing farm tree species such as *Pongamia pinnata, Grewia optiva* etc. which have been cultivated for agricultural inputs to farms but have not been of interest in commercial forestry. In spite of being an incomplete list of fast growing indigenous trees, the forest plantation data on yields adequately reveals that Eucalyptus is among the slower growing species even for woody biomass production. The Eucalyptus hybrid, the most dominantly planted Eucalyptus species, has different growth rates at different ages and on different sites as shown in Table 3.

The points that emerge from Tables 2 and 3 are:

i)   In terms of yields measured as mean annual increment
      (MAI) Eucalyptus is a slow producer of woody biomass
      even under very good soil conditions and water availabil-
      ity.
ii)   When the site is of poor quality such as eroded soils or barren
      land Eucalyptus yields are insignificant.
iii) The growth rate of Eucalyptus under the best conditions is
      not uniform for different age groups. It falls very drastically
      after 5 or 6 years.

Table 2		
Some indigenous species which are comparatively fast-growing		
Name of Species	Age (Yrs)	MAI (CuM/ha)
Duabanga sonneratioides	47	19
Alnus nepalensis	22	16
Terminalia myriocarpa	8	15
Evodia meliafolia	11	10
Michelia Champaca	8	18
Lophopetalum fibriatum	17	15
Casuarina equisetifolia	5	15
Shorea robusta	30	11
Toona ciliata	5	19
Trewia nudiflora	13	13
Artocarpus chaplasha	10	16
Dalbergia sissoo	11	34
Gmelina arborea	3	22
Tectona grandis	10	12
Michelia oblonga	14	18
Bischofia javanica	7	13
Broussonatia papyrifera	10	25
Bucklandia populnea	15	9
Terminalia tomentosa	4	10
Kydia calycina	10	11

Table 3			
Yield Table for Eucalyptus hybrid			
Site Quality	Age	MAI CuM/ha (OB)	Current AI CuM/ha (OB)
Good	3	8.1	–
	4	11.3	10.6
	5	13.5	22.3
	6	14.4	18.7
	7	13.9	11.3
	8	13.5	10.6
	9	12.9	8.0
	10	12.3	6.7
	11	11.6	5.2
	12	11.0	3.5
	13	10.4	3.6
	14	9.9	3.7
	15	9.4	1.9
Poor	3	0.1	–
	4	0.4	1.4
	5	0.7	1.7
	6	0.8	1.7
	7	0.9	1.2
	8	1.0	1.4
	9	1.0	1.0
	10	1.0	1.3
	11	1.0	1.1
	12	1.2	0.7
	13	1.0	0.8
	14	0.9	0.8
	15	0.9	0.4
OB = Over Bark; MAI = Mean Annual Increment			

Scientific evidence on the biomass productivity does not support the claim that the Eucalyptus is faster growing than other alternative species or that it grows well even on degraded lands. Under rainfed conditions the best yields achieved for Eucalyptus has been 10 tons/ha/yr. On the other hand:

According to Dr K S Rao and Dr K K Bokil (unpublished reports) one hectare of Prosopis yields 31 tons of bone dry firewood per year. At Vatva in Ahmedabad district, Gujarat state, annual production of firewood from Prosopis was recorded as 25 tons/ha/yr under rainfed conditions.

A comparison of growth rate of 10 species by the Gujarat Forest Department shows that the Eucalyptus emerges at the bottom of the list. The Eucalyptus, quite clearly, will not fill the gap in the demand of woody biomass more effectively than other faster growing species which are also better adapted to the Indian conditions.

Forests and trees have been producing various kinds of biomass, satisfying diverse human needs. Modern forestry management, however, came as a response to the demands for woody biomass for commercial and industrial purposes. The growth rate of the species that is provided by modern forestry is, therefore, restricted in two ways. Firstly, it is restricted to the increment and growth of the trunk biomass alone. Even in this restricted spectrum, Eucalyptus ranks very low in terms of growth and biomass productivity.

Human needs for biomass are, however, not restricted to the consumption and use of woody biomass alone. The maintenance of life support systems is a function performed mainly by the crown biomass of trees. It is this component of trees that can contribute positively towards the maintenance of the hydrological and nutrient cycles. It is also the most important source for the production of biomass for consumption as fuel, fodder, manure, fruits etc. Social forestry as distinct from commercial forestry, to which it is supposed to be a corrective, is in principle, aimed at the maximisation of the production of all types of useful biomass which improve ecological stability and satisfy diverse and basic biomass needs. The appropriate unit of assessment of growth and yields of different tree species for social forestry programmes cannot be restricted to the woody biomass production for commercial use. It must, instead, be specific to the end use of biomass. The crisis in biomass for animal feed, quite evidently, cannot be overcome by planting trees that are fast growing from the perspective of the pulp industry, but are absolutely unproductive as far as fodder requirements are concerned.

The assessment of yields in social forestry must include the diverse types of biomass which provides inputs to the agro-ecosystem. When the objective for tree planting is the production of fodder or green fertilizer, it is relevant to measure crown biomass productivity. India, with its rich genetic diversity in plants and animals, is richly endowed with various types of fodder trees which have annual yields of crown biomass that is much higher than the total biomass produced by Eucalyptus plantations as indicated in Table 4.

Table 4	
Crown biomass productivity of some well-known fodder trees	
Name of species	Crown Biomass Tons/ha/yr
Acacia nilotica	13 - 27
Grewia optiva	33
Bauhinia	47
Ficus	17.5
Leucena leucocephala	07.5
Morus alba	24
Prosopis sineraria	30

An important biomass output of trees that is never assessed by foresters who look for timber and wood is the yield of seeds and fruits. Fruits trees such as jack, jaman, mango, tamarind etc. have been important components of indigenous forms of social forestry as practised over centuries in India. After a brief gestation period fruit trees yield annual harvests of edible biomass on a sustainable and renewable basis.

Tamarind trees yield fruits for even over two centuries. Other trees, such as, neem, pongamia and sal provide annual harvest of seeds which yield valuable non-edible oils. These diverse yields of biomass provide important sources of livelihood for millions of tribes or rural people. The coconut, for example, besides providing fruits and oil, provides leaves used in thatching huts and supports the large coir industry in the country. Since social forestry pro-grammes in their present form have been based on only the knowledge of foresters who have been trained only to look for the woody biomass in the tree, these important high yielding species of other forms of biomass have been totally ignored in these programmes. Two species on which ancient farm forestry systems in arid zones have laid special stress are pongmia and tamarind. Both these trees are multi-dimensional producers of firewood, fertiliser, fodder, fruit and oil seed. More significantly, components of the crown biomass that are harvested from fruit and fodder trees leave the living tree standing to perform its essential ecological functions in soil and water conservation. In contrast, the biomass of the Eucalyptus is useful only after the tree is felled.

Figures 2 and 3 describe the comparative biomass contribution of indigenous trees and Eucalyptus. Afforestation strategies based dominantly on Eucalyptus are not therefore, the most effective mechanism for tiding over the serious biomass crisis facing the country. The benefits of Eucalyptus have often been unduly exaggerated through the myth of its fast growth and high yields. The myth has become pervasive because of the unscientific and

Figure 2: The Contribution of Traditional Tree Species to the Rural Life-support Systems.

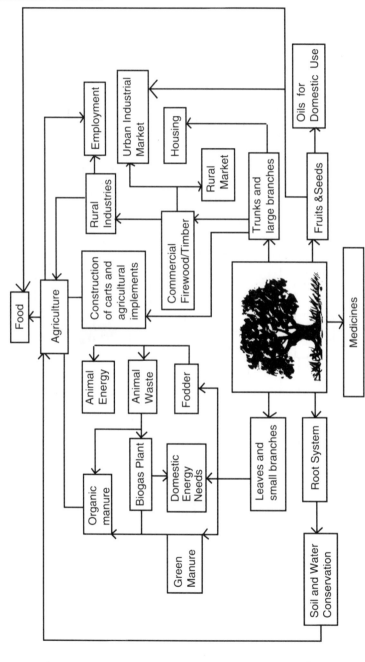

Source: *Shiva et al (1981)*

Figure 3: The Comparative Contribution of Eucalyptus to the Rural Life-support Systems

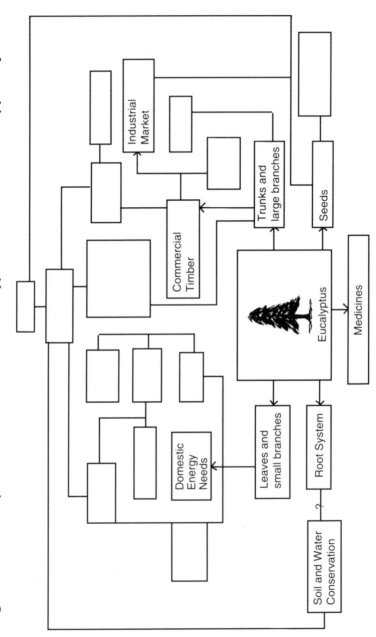

unjustified advertisement of the species. It has also been aided by the linear growth of Eucalyptus in one dimension while most indigenous trees have broad crowns that grow in three dimensions.

The Green Revolution and 'Miracle' Seeds

In agriculture, too, the monoculture mind creates the monoculture crop. The miracle of the new seeds has most often been communicated through the term 'high-yielding varieties' (HYV). The HYV category is a central category of the Green Revolution paradigm. Unlike what the term suggests, there is no neutral or objective measure of 'yield' on the basis which the cropping systems based on miracle seeds can be established to be higher yielding than the cropping systems they replace. It is now commonly accepted that even in the most rigorous of scientific disciplines such as physics, there are no neutral observational terms. All terms are theory laden.

The HYV category is similarly not a neutral observational concept. Its meaning and measure is determined by the theory and paradigm of the Green Revolution. And this meaning is not easily and directly translatable for comparison with the agricultural concept of indigenous farming systems for a number of reasons. The Green Revolution category of HYV is essentially a reductionist category which decontextualises contextual properties of both the native and the new varieties. Through the process of decontextualisation, costs and impacts are externalised and systemic comparison with alternatives is precluded.

Cropping systems, in general, involve an interaction between soil, water and plant genetic resources. In indigenous agriculture, for example, cropping systems include a symbiotic relationship between soil, water, farm animals and plants. Green Revolution agriculture replaces this integration at the level of the farm with the integration of inputs such as seeds and chemicals. The seed/ chemical package sets up its own interactions with soils and water systems, which are, however, not taken into account on the assessment of yields.

Modern plant breeding concepts like HYVs reduce farming systems to individual crops and parts of crops (Figure 4). Crop components of one system are then measured with crop components of another. Since the Green Revolution strategy is aimed at increasing the output of a single component of a farm, at the cost of decreasing other components and increasing external inputs, such a partial comparison is by definition biased to make the new varieties 'high yielding' although when at the systems level, they may not be.

Traditional farming systems are based on mixed and rotational cropping systems of cereals, pulses, oilseeds with diverse varieties of each crop, while the Green Revolution package is based on genetically uniform monocultures. No realistic assessments are ever made of the yield of the diverse crop outputs in the mixed and rotational systems. Usually the yield of a single crop like wheat or maize is singled out and compared to yields of new varieties. Even if the yields of all the crops were included, it is difficult to convert a measure of pulse into an equivalent measure of wheat, for example, because in the diet and in the ecosystem, they have distinctive functions.

The protein value of pulses and the calorie value of cereals are both essential for a balanced diet, but in different ways and one cannot replace the other as illustrated in Table 1. Similarly, the nitrogen fixing capacity of pulses is an invisible ecological contribution to the yield of associated cereals. The complex and diverse cropping systems based on indigenous varieties are therefore not easy to compare to the simplified monocultures of HYV seeds. Such a comparison has to involve entire systems and

Figure 4: How The Green Revolution Makes Unfair Comparisons

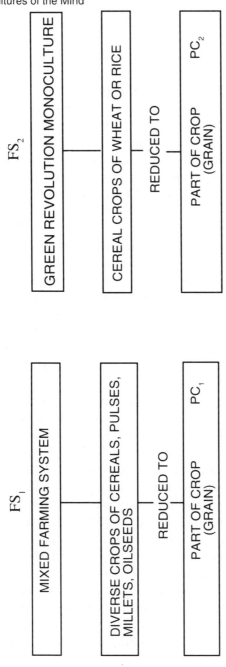

FS_1

MIXED FARMING SYSTEM

DIVERSE CROPS OF CEREALS, PULSES, MILLETS, OILSEEDS

REDUCED TO

PART OF CROP (GRAIN) PC_1

FS_2

GREEN REVOLUTION MONOCULTURE

CEREAL CROPS OF WHEAT OR RICE

REDUCED TO

PART OF CROP (GRAIN) PC_2

▲ The real scientific comparison should be between two farming systems - FS_1 and FS_2, with the full range of inputs and outputs included.

▲ This would be the comparison if FS_2 was not given immunity from an ecological evaluation.

▲ In the Green Revolution strategy, a false comparison is made between PC_1 and PC_2.

▲ So while PC_2 > PC_1 generally FS_1 > FS_2

Source: *Shiva (1989)*

cannot be reduced to a comparison of a fragment of the farm system. In traditional-farming systems, production has also involved maintaining the conditions of productivity. The measurement of yields and productivity in the Green Revolution paradigm is divorced from seeing how the processes of increasing output affect the processes that sustain the condition for agricultural production. While these reductionist categories of yield and productivity allow a higher destruction that affects future yields, they also exclude the perception of how the two systems differ dramatically in terms of inputs (Figure 5).

The indigenous cropping systems are based only on internal organic inputs. Seeds come from the farm, soil fertility comes from the farm and pest control is built into the crop mixtures. In the Green Revolution package, yields are intimately tied to purchased inputs of seeds, chemical fertilisers, pesticides, petroleum and to intensive and accurate irrigation. High yields are not intrinsic to the seeds, but are a function of the availability of required inputs, which in turn have ecologically destructive impacts (Figure 6).

As Dr Palmer concluded in the United Nations Research Institute for Social Development's 15 nation study of the impact of the seeds, the term 'high yielding varieties' is a misnomer because it implies that the new seeds are high-yielding in and of themselves. The distinguishing feature of the seeds, however, is that they are highly responsive to certain key inputs such as fertilisers and irrigation. Palmer therefore suggested the term 'high-responsive varieties' (HRVs) in place of 'high yielding varieties'.[25] In the absence of additional inputs of fertilisers and irrigation, the new seeds perform worse than indigenous varieties. With the additional inputs, the gain in output is insignificant compared to the increase in inputs. The measurement of output is also biased by restricting it to the marketable part of crops. However, in a country like India, crops have traditionally been bred and cultivated to produce not just food for man but fodder for animals, and organic

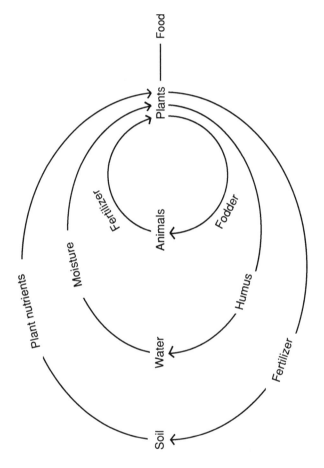

Figure 5: Internal Input Farming System

Source: Shiva (1989)

Figure 6: External Input Farming System

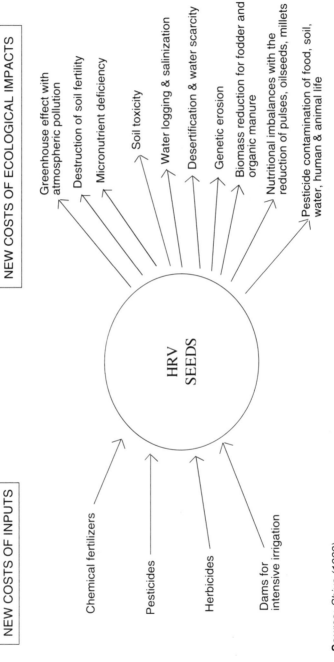

Source: *Shiva (1989)*

fertiliser for soils. According to A K Yegna Narayan Aiyer, a leading authority on agriculture, 'as an important fodder for cattle and in fact as the sole fodder in many tracts, the quantity of straw obtainable per acre is important in this country. Some varieties which are good yielders of grains suffer from the drawback of being low in respect to straw'.[26] He illustrated the variation in the grain-straw ratio with yields from the Hebbal farm.

In the breeding strategy for the Green Revolution, multiple uses of plant biomass seem to have been consciously sacrificed for a single use, with non-sustainable consumption of fertiliser and water. The increase in marketable output of grain has been achieved at the cost of decrease of biomass for animals and soils and the decrease of ecosystem productivity due to over-use of resources.

The increase in production of grain for marketing was achieved in the Green Revolution strategy by reducing the biomass for

Table 5		
Grain and Straw Production of Rice Varieties		
Name of Variety	Grain (in lb. per acre)	Straw (in lb. per acre)
Chintamani sanna	1,663	3,333
Budume	1,820	2,430
Halubbalu	1,700	2,740
Gidda Byra	1,595	2,850
Chandragutti	2,424	3,580
Putta Bhatta	1,695	3,120
Kavada Bhatta	2,150	2,940
Garike Sanna	2,065	2,300
Alur sanna	1,220	3,580
Bangarkaddi	1,420	1,760
Banku (rainy season 1925-26)	1,540	1,700
G.E.B. - do -	1,900	1,540

internal use on the farm. This is explicit in a statement by Swaminathan:

> *High yielding varieties of wheat and rice are high yielding because they can use efficiently larger quantities of nutrients and water than the earlier strains, which tended to lodge or fall down if grown in soils with good fertility... They thus have a 'harvest index' (i.e. the ratio of the economic yield to the total biological yield) which is more favourable to man. In other words, if a high yielding strain and an earlier tall variety of wheat both produce, under a given set of conditions, 1000 kg of dry matter, the high yielding strain may partition this dry matter into 500 kg for grain and 500 kg for straw. The tall variety, on the other hand, may divert 300 kg for grain and 700 kg for straw.*[27]

The reduction of outputs of biomass for straw production was probably not considered a serious cost since chemical fertilisers were viewed as a total substitute for organic manure, and mechanisation was viewed as a substitute to animal power. According to one author,

> *It is believed that the 'Green Revolution' type of technological change permits higher grain production by changing the grain foliage ratio... At a time there is urgency for increasing grain production, an engineering approach to altering the product mix on an individual plant may be advisable, even inevitable. This may be considered another type of survival technological change. It uses more resources, returns to which are perhaps unchanged (if not diminished).*[28]

It was thus recognised that in terms of over-all plant biomass, the Green Revolution varieties could even reduce the overall yields of crops and create scarcity in terms of output such as fodder.

Finally, there is now increasing evidence that indigenous varieties could also be high yielding, given the required inputs.

Richaria has made a significant contribution to the recognition that peasants have been breeding high yielding varieties over centuries. Richaria reports:

A recent varietal-cum-agronomic survey has shown that nearly 9% of the total varieties grown in U P fall under the category of high yielding types (3,705 kgs and above per hectare).

A farmer planting a rice variety called Mokdo of Bastar who adopted his own cultivation practices obtained about 3,700 to 4,700 kgs of paddy per hectare. Another rice grower of Dhamtari block (Raipur) with just a hectare of rice land, falling not in an uncommon category of farmers, told me that he obtains about 4,400 kgs of paddy per hectare from Chinnar variety, a renowned scented type, year after year with little fluctuations. He used FYM supplemented at times with a low dose of nitrogen fertilizers. For low lying area in Farasgaon Block (Bastar) a non-lodging tall rice variety Surja with bold grains and mildly scented rice may compete with Jaya in yield potential at lower doses of fertilization, according to a local grower who showed me his crop of Surja recently.

During my recent visit of the Bastar area in the middle of November 1975 when the harvesting of a new rice crop was in full swing in a locality, in one of the holdings of an adivasi cultivator, Baldeo of Bhatra tribe in village Dhikonga of Jugalpur block, I observed a field of Assam Chudi ready for harvest with which the adivasi cultivator has stood for crop competition. The cultivator has applied the fertilizer approximately equal to 50 kg/N ha and has used no plant protection measures. He expected a yield of about 5,000 kg/ha. These are good cases of applications of an intermediate technology for increasing rice production. The yields obtained by those farmers fall in or above the minimum limits set for high yields and these methods of cultivation deserve full attention.[29]

India is a Vavilov centre or centre of genetic diversity of rice. Out of this amazing diversity, Indian peasants and tribals have selected and improved many indigenous high yielding varieties.

In South India, in semi-arid tracts of the Deccan, yields went up to 5,000 kg/ha under tank and well irrigation. Under intensive manuring, they could go even higher. As Yegna Narayan Aiyer reports:

The possibility of obtaining phenomenal and almost unbelievably high yields of paddy in India has been established as the result of the crop competitions organised by the Central Government and conducted in all states. Thus even the lowest yield in these competitions has been about 5,300 lbs/acre, 6,200 lbs/acre in West Bengal, 6,100, 7,950, and 8,258 lbs/ acre in Thirunelveli, 6,368 and 7,666 kg/ha in South Arcot, 11,000 lbs/ acre in Coorg and 12,000 lbs/acre in Salem.[30]

The Green Revolution package was built on the displacement of genetic diversity at two levels. Firstly, mixtures and rotation of diverse crops like wheat, maize, millets, pulses, and oil seeds were replaced by monocultures of wheat and rice. Secondly, the introduced wheat and rice varieties reproduced over large-scale as monocultures came from a very narrow genetic base, compared to the high genetic variability in the population of traditional wheat or rice plants. When 'HYV' seeds replace native cropping systems diversity is lost and is irreplaceable.

The destruction of diversity and the creation of uniformity simultaneously involves the destruction of stability and the creation of vulnerability. Local knowledge on the other hand, focuses on multiple-use of diversity. Rice is not just grain, it provides straw for thatching and mat-making, fodder for livestock, bran for fish ponds, husk for fuel. Local varieties of crops are selected to satisfy these multiple uses. The so-called HYV varieties increase grain production, by decreasing all other outputs, increasing external inputs, and introducing ecologically destructive impacts.

Local knowledge systems have evolved tall varieties of rice and wheat to satisfy multiple needs. They have evolved sweet

Cassava varieties whose leaves are palatable as fresh greens. However, all dominant research on cassava has focused on breeding new varieties for tuber yields, with leaves which are unpalatable.

Ironically, breeding for a **reduction** in usefulness has been viewed as **important** in agriculture, because uses outside those that serve the market are not perceived and taken into account. The new ecological costs are also left out as 'externalities' thus rendering an inefficient wasteful system productive.

There is, moreover, a cultural bias, which favours the modern system, a bias which becomes evident in the naming of plant varieties. The indigenous varieties, or land races, evolved through both natural and human selection, and produced and used by Third World farmers world-wide are called 'promotive cultivar'. Those varieties created by modern plant breeders in international agricultural research centres or by transnational seed corporations are called 'advanced' or 'elite'.

Yet the only aspect in which the new varieties have really been 'advanced' has been in their ecologically appropriate systems not through test and evaluation, but through the unscientific rejection of local knowledge as primitive and the false promise of 'miracles' – of 'miracle' trees and 'miracle' seeds. Yet as Angus Wright has observed:

One way in which agricultural research went wrong was precisely in saying and allowing it to be said that some miracle was being produced.... Historically, science and technology made their first advances by rejecting the idea of miracles in the natural world. Perhaps it would be best to return to that position.[31]

The Non-Sustainability of Monocultures

The crucial characteristic of monocultures is that they do not merely displace alternatives, they destroy their own basis. They are neither tolerant of other systems, nor are they able to reproduce themselves sustainably. The uniformity of the 'normal' forest that 'scientific' forestry attempts to create becomes a prescription for non-sustainability.

The displacement of local forest knowledge by 'scientific' forestry was simultaneously a displacement of the forest diversity and its substitution by uniform monocultures. Since the biological productivity of the forest is ecologically based on its diversity, the destruction of local knowledge, and with it of plant diversity, leads to a degradation of the forest and an undermining of its sustainability. The increase in productivity from the commercial point of view destroys productivity from the perspective of local communities. The uniformity of the managed forest is meant to generate 'sustained yields'. However, uniformity destroys the conditions of renewability of forest eco-systems, and is ecologically non-sustainable.

In the commercial forestry paradigm 'sustainability' is a matter of supply to the market, not the reproduction of an ecosystem in its biological diversity or hydrological and climatic stability. As Schlich states, 'forest working plans regulate, according to time and locality, the management of forests in such a manner, that the objects of the industry are as full as possible realised.'[32] Sustained yield management is aimed at producing 'the best financial results, or the greatest volume, or the most suitable class of produce'. If this could be ensured while maintaining the forest eco-system, we would have sustainability in nature, not just short-term sustainability for market supplies of industrial and commercial wood. However, 'sustained yields' as conceived in forestry management, is based on the assumption that the real forest, or the natural forest is not a 'normal' forest, it is an 'abnormal' forest. When 'normalcy' is determined by the demands of the market, the non-marketable components of the natural forest ecosystem are seen as 'abnormal' and are destroyed by prescriptions of forest working plans.

Uniformity in the forest is the demand of centralised markets and centralised industry. However, uniformity acts against nature's processes. The transformation of mixed natural forests into uniform monocultures allows the direct entry of tropical sun and rain, baking the forest soils dry in the heat, washing the soils off in the rain. Less humid conditions are the reason for rapid retrogression of forest regions. The recent fires of Kalimantan are largely related to the aridisation caused by the conversion of rainforests into plantations of Eucalyptus and Acacias. Floods and drought are created where the tropical forest had earlier cushioned the discharge of water.

In tropical forests, selective felling of commercial species produces only small yields $(5-25\,m^3/ha)$ whereas clearfelling might produce as much as $450m^3/ha$. The non-sustainability of selection fellings is also borne out by the experience of PICOP, a joint venture set up in 1952, between the American firm, International Paper Company, the world's largest paper producer, and the Andre Soriano Corporation in the Philippines. The company takes only about 10% of the total volume of wood, roughly 73 cubic yards per acre of virgin forest. But the company's measurements of annual growth show that the second rotation will only yield 37 cubic yards of useful wood per acre, half as much as the first cut, and not enough to keep the company's plywood, veneer, and sawmills functioning at a profitable level.

'Sustained yields' can be managed for PICOP by reducing the diameter for extraction. At present the government allows PICOP to take out all trees larger than 32 inches in diameter, and a certain proportion of those that are 24 inches or more in diameter. If on the second rotation they could take out all trees bigger than 12 or 16 inches around, they could sustain supplies for another rotation. Taking smaller trees on the second cut would not, of course, make the forest grow faster, for a third, fourth, and fifth rotation.

PICOP's plantations have also failed. It had to replant 30,000 acres of a variety of Eucalyptus from Papua New Guinea that was attacked by pests. Its pine plantations of 25,000 acres have also failed. At $400 per acre, that was a $10 million mistake.

Angel Alcala, Prof. of Biology of Siliman University in the Philippines observes that selective logging is good in theory, but it does not really work. 'With selective logging, you are supposed to take only a few trees and leave the rest to grow, so you can return later and take some more, without destroying the forest. This is supposed to be a sustainable system. But here, although they use the phrase selective logging, there is only one harvest, a big one. After that no more.'[33]

One study found that 14% of a logging area is cleared for roads, and another 27% for skidder trucks. Thus more than 40% of a concession can be stripped of protective vegetation and highly liable to erosion. It can be as high as 60%.[34]

In dipterocarp forests, with an average of 58 trees per acre, for every 10 that are deliberately felled, 13 more are broken or damaged. Selective loggers damage more trees than they harvest. In one Malaysian dipterocarp forest, only 10% of the trees were harvested, 55% were destroyed or severely damaged. Only 33% were unharmed. In Indonesia, according to the manager of Georgia-Pacific, they damage or destroy more than three times as many as they deliberately harvest.[35]

According to the UNESCO report on tropical forest ecosystems, not many forests are rich enough to allow true selective working – the removal of each tree (of desirable species) as soon as it reaches commercial size. Not only does each tree cause considerable damage when it falls, but the heavy logging equipment needed causes further damage. To sum up, true selective felling is impracticable regardless of the structure, composition and dynamism of the original stands.

This paradigm which destroys the diversity of the forest community either by clear felling or selective felling simultaneously destroys the very **conditions** for the renewal of the forest community. While species diversity is what makes the tropical forest biologically rich, and sustainable, this same diversity leads to allow density of individual species. The reductionist paradigm thus converts a biologically rich system into an impoverished resource and hence a non-renewable one. Thus while the annual biological production of tropical broadleaved forest is 300 tons/ha compared to 150 tons/ha, the annual production of commercial wood is only 0.14 m^3/ha on the average in tropical forests compared to 1.08 m^3. In tropical Asia, commercial production is 0.39 m^3/ha due to the richness in diversity of commercial species of the dipterocarp forests.[36]

In the dominant system, financial survival strategies determine the concept of 'sustained yield', which are in total violation of the principles of sustaining biological productivity. Sustained yields based on continuously reducing exploitable diameter classes leads to biological suicide, and a total destruction of forests.

Fahser reports how a forestry project in Brazil, aimed at 'self-help' and satisfying basic needs, destroyed both the forests and the communities whose improvement it was aimed at:

With the building up of the first Faculty of Forestry Science and the imparting of modern forestry knowledge, a milestone was actually reached in the forests of Brazil. A greater knowledge of economics encouraged trained people towards new approaches; the natural forest with its many species was replaced by huge timber plantations of fir and eucalyptus; weak and unreliable human workers were replaced by powerful timber harvesting machinery; the hitherto untouched coastal mountain ranges were conquered, using rope cranes as an elegant means of transport.

Since forestry development aid began, afforestation in Parana has dropped from about 40% to its present level of 8%. Transformation into steppe, erosion and periodical flooding are on the increase. Our highly qualified Brazilian counterparts are now shifting their interest to the Amazon regions of the north where there are still plenty of forests and where they are 'managing' cellulose timber plantations (eg of Gmelina arborea) with rotation periods of only six years.

What happened to the population during the roughly 20 years period of the project, to those people whose basic needs were to be satisfied and who were to be given aid so that they could help themselves? Parana is now largely cleared of forest and full of mechanised agriculture. Most Indios and many immigrants who lived there at subsistence level or as small farmers have silently disappeared, become impoverished and collected in the slums (favelas) in the vicinity of the cities. In forestry the capital-intensive unit on the mechanisation pattern of north America and Scandinavia is now dominant. Only a few experts and a few wage-earners are still needed for peak work periods.[37]

Where the local knowledge is not totally extinct, communities resist the ecological destruction of introduced monocultures. 'Greening' with eucalyptus works against nature and its cycles, and it is being resisted by communities who depend on the stability of nature's cycles to provide sustenance in the form of food and water. The Eucalyptus guzzles nutrients and water and, in the specific conditions of low rainfall zones, gives nothing back but terpenes to the soil. These inhibit the growth of other plants and are toxic to soil organisms which are responsible for building soil fertility and improving soil structure. The eucalyptus certainly increased cash and commodity flows, but it resulted in a disastrous interruption of organic matter and water flows within the local ecosystem. Its proponents failed to calculate the costs in terms of the destruction of life in the soil, the depletion of water resources and the scarcity of food and fodder that eucalyptus cultivation creates. Nor did they, while trying to shorten rotations for harvesting, see that tamarind, jackfruit and honge have very short rotations of one year in which the biomass harvested is far higher than that of eucalyptus, which they nevertheless declared a 'miracle' tree. The crux of the matter is that fruit production was never the concern of forestry in the reductionist paradigm – it focused on wood, and wood for the market, alone. Eucalyptus as an exotic, introduced in total disregard of its ecological appropriateness has thus become an exemplar of anti-life afforestation.[38]

People everywhere have resisted the expansion of eucalyptus because of its destruction of water, soil and food systems. On 10 August 1983, the small peasants of Barha and Holahalli villages in Tumkur district (Karnataka) marched en masse to the forestry nursery and pulled out millions of eucalyptus seedlings, planting tamarind and mango seeds in their place. This gesture of protest, for which they were arrested, spoke out against the virtual planned destruction of soil and water systems by eucalyptus cultivation. It also challenged the domination of a forestry-science that had reduced all species to one (the eucalyptus), all needs to one, (that of the pulp industry), and all knowledge to one (that of the World Bank and forest officials). It challenged the myth of the miracle tree: tamarind and mango are symbols of the energies of nature and of local people, of the links between these seeds and the soil, and of the needs that these trees – and others like them – satisfy in keeping the earth and the people alive. Forestry for food – food for the soil, for farm animals, for people – all women's and peasants' struggles revolve around this theme, whether in Garhwal or

Karnataka, in the Santhal Perganas or Chattisgarh, in reserved forests, farmlands or commons. In June 1988, in protest against eucalyptus planting, villagers in northern Thailand burned down eucalyptus nurseries at a forestry station.

The destruction of diversity in agriculture has also been a source of non-sustainability. The 'miracle' varieties displaced the traditionally grown crops and through the erosion of diversity, the new seeds became a mechanism for introducing and fostering pests. Indigenous varieties, or land races are resistant to locally occurring pests and diseases. Even if certain diseases occur, some of the strains may be susceptible, while others will have the resistance to survive. Crop rotations also help in pest control. Since many pests are specific in particular plants, planting crops in different seasons and different years causes large reductions in pest population. On the other hand, planting the same crop over large areas year after year encourages pest build ups. Cropping systems based on diversity thus have a built-in protection.

Having destroyed nature's mechanisms for controlling pests through the destruction of diversity, the 'miracle' seeds of the Green Revolution became mechanisms for breeding new pests and creating new diseases. The treadmill of breeding new varieties runs incessantly as ecologically vulnerable varieties create new pests which create the need for breeding yet newer varieties.

The only miracle that seems to have been achieved with the breeding strategy of the Green Revolution is the creation of new pest and diseases, and with them the ever increasing demand for pesticides. Yet the new costs of new pests and poisonous pesticides were never counted as part of the 'miracle' of the new seeds that modern plant breeders had given the world in the name of increasing 'food security'.

The 'miracle seeds' of the Green Revolution were meant to free the Indian farmer from constraints imposed by nature. Instead, large scale monocultures of exotic varieties generated a new ecological vulnerability by reducing genetic diversity and destabilising soil and water systems. The Green Revolution led to a shift from earlier rotations of cereals, oilseeds, and pulses to a paddy-wheat rotation with intensive inputs of irrigation and chemicals. The paddy-wheat rotation has created an ecological backlash with serious problems of waterlogging in canal-irrigated regions and groundwater mining in tubewell irrigated regions. Further, the high yielding varieties have led to large-scale micronutrient deficiencies in soils, particularly iron in paddy cultivation and manganese in wheat.

These problems were built into the ecology of the HYV's even though they were not anticipated. The high water demands of these seeds necessitated high water inputs, and hence the hazards of desertification through water logging in some regions and desertification and aridisation in others. The high nutrient de-mands caused micronutrient deficiencies on the one hand, but were also unsustainable because increased applications of chemi-cal fertilizers were needed to maintain yields, thus increasing costs without increasing returns. The demand of the HYV seeds for intensive and uniform inputs of water and chemicals also made large-scale monocultures an imperative, and monocultures being highly vulnerable to pests and diseases, a new cost was created for pesticide applications. The ecological instability inherent in HYV seeds was thus translated into economic non-viability. The mira-cle seeds were not such a miracle after all.

Sustainable agriculture is based on the recycling of soil nutri-ents. This involves returning to the soil, part of the nutrients that come from the soil either directly as organic fertilizer, or indirectly through the manure from farm animals. Maintenance of the nutrient cycle, and through it the fertility of the soil, is based on this inviolable law of return, which is a timeless, essential element of sustainable agriculture.

The Green Revolution paradigm substituted the nutrient cycle with linear flows of purchased inputs of chemical fertilizers from factories and marketed outputs of agricultural commodities. Yet, the fertility of soils cannot be reduced to NPK in factories, and agriculture productivity necessarily includes returning to the soil part of the biological products that the soil yields. Technologies cannot substitute nature and work outside nature's ecological processes without destroying the very basis of production. Nor can markets provide the only measure of 'output' and 'yields'.

The Green Revolution created the perception that soil fertility is produced in chemical factories, and agricultural yields are measured only through marketed commodities. Nitrogen fixing crops like pulses were therefore displaced. Millets which have high yields from the perspective of returning organic matter to the soil, were rejected as 'marginal' crops. Biological products not sold on the market but used as internal inputs for maintaining soil fertility were totally ignored in the cost-benefit equations of the Green Revolution miracle. They did not appear in the list of inputs because they were not purchased, and they did not appear as outputs because they were not sold.

Yet what is 'unproductive' and 'waste' in the commercial context of the Green Revolution is now emerging as productive in the ecological context and as the only route to sustainable agriculture. By treating essential organic inputs that maintain the integrity of nature as 'waste', the Green Revolution strategy ensured that fertile and productive soils are actually laid waste. The 'land-augmenting' technology has proved to be a land-degrading and land-destroying technology. With the greenhouse effect and global warming, a new dimension has been added to the ecologically destructive effect of chemical fertilizers. Nitrogen-based fertilizers release nitrous oxide to the atmosphere which is one of the greenhouse gases causing global warming. Chemical farming has thus contributed to the erosion of food security through the pollution of land, water and the atmosphere.[39]

Democratising Knowledge

Modern silviculture as an exclusivist knowledge system which focuses exclusively on industrial wood production displaces local knowledge systems which view the forest in the perspective of food production, fodder production and water production. The exclusive focus on industrial wood destroys the food, fodder and water production capacities of the forest. It disrupts links between forestry and agriculture, and in attempting to increase commercial/industrial wood, it creates a monoculture of tree species. The eucalyptus has become a symbol of this monoculture.

Modern agriculture focuses exclusively on agricultural commodity production. It displaces local knowledge systems which view agriculture as the production of diverse food crops with internal inputs, and replaces it with monocultures of introduced varieties needing external industrial inputs. The exclusive focus on external inputs and commercial outputs, destroys diverse food crops such as pulses, oilseeds and millets, disrupts the local ecological cycles; and in attempting to increase single crop output, it creates monocultures of crop varieties. The HYV becomes a symbol of this monoculture.

The crises of the dominant knowledge system has many facets,

(a)   Since dominant knowledge is deeply wedded to economism, it is unrelated to human needs. Ninety percent of such production of knowledge could be stopped without any risk of human deprivation. On the contrary, since a large part of such knowledge is a source of hazards, and threats to human life (Bhopal, Chernobyl, Sandoz) its end would improve the possibilities of human well being.

(b)   The political implications of the dominant knowledge system are inconsistent with equality and justice. It is disrupting of cohesion within local communities and polarises society into those with access and those without it, both in respect to the knowledge systems and the power system.

(c)   Being inherently fragmenting and having built in obsolescence, dominant knowledge creates an alienation of wisdom from knowledge and dispenses with the former.

(d)   It is inherently colonising, inherently mystifying, shielding colonisation by mystification.

(e)   It breaks away from concrete contexts, disqualifying as inadequate the local and concrete knowledge.

(f)   It closes access and participation to a plurality of actors.

(g)   It leaves out a plurality of paths to knowing nature and the universe. It is a monoculture of the mind.

Modern western knowledge is a particular cultural system with a particular relationship to power. It has, however, been projected as above and beyond culture and politics. Its relationship with the project of economic development has been invisible; and therefore it has become a more effective legitimiser for the homogenisation of the world and the erosion of its ecological and cultural richness. The tyranny and hierarchy privileges that are part of the development drive are also part of the globalising knowledge in which the development paradigm is rooted and from which it derives its rationalisation and legitimisation. The power by which the dominant knowledge system has subjugated all others makes it exclusive and undemocratic.

Democratising of knowledge becomes a central precondition for human liberation because the contemporary knowledge system excludes the humane by its very structure. Such a process of democratisation would involve a redefining of knowledge such

Comparison of Local and Dominant Knowledge Systems	
Local System	**Dominant System**
1. forestry and agriculture integrated	forestry separate from agriculture
2. Integrated systems have multidimensional outputs. Forests produce wood, food, fodder, water etc. Agriculture produces diversity of food crops.	Each separate system made one dimensional. Forests produce only commercial wood. Agriculture produces only commercial crops with industrial inputs.
3. Productivity in local system is a multi-dimensional measure, which has a conservation aspect.	Productivity is a one dimensional measure which is unrelated to conservation.
4. Increasing productivity in these knowledge systems involves increasing the multidimensional outputs, and strengthening the integration.	Increasing productivity in these knowledge systems involves increasing one dimensional output by breaking up integrations and displacing diverse outputs.
5. Productivity based on conservation of diversity.	Productivity based on creation of monocultures and destruction of diversity
6. Sustainable system	Non-sustainable system.

that the local and diverse become legitimate as knowledge, and they are viewed as indispensable knowledge because concreteness is the reality, and globalisation and universalisation are more mere abstractions which have violated the concrete and hence the real. Such a shift from the globalising to the local knowledge is important to the project of human freedom because it frees knowledge from the dependency on established regimes of thought, making it simultaneously more autonomous and more authentic. Democratisation based on such an 'insurrection of subjugated knowledge' is both a desirable and necessary component of the larger processes of democratisation because the earlier paradigm is in crisis and in spite of its power to manipulate, is unable to protect both nature and human survival.

References

1.   Harding, S. 1986, *The Science Question in Feminism*, Cornell University Press, Ithaca, p8

2.   Kuhn, T. 1972, *The Structure of Scientific Revolutions*, University of Chicago Press, Chicago

3.   Horton, R. 1967, African Traditional Thought and Western Science, *Africa* 37, 2

4.   Harding, op cit p30

5.   Shiva, V. *Ecology and the Politics of Survival*, UNU Tokyo and Sage, New Delhi, London, Newbury Park

6.   Caufield, C. 1986, *In the Rainforest*, Picador, London, p60

7.   Hong, E. 1987, *Natives of Sarawak*, Institut Masyarakat, Malaysia p137

8.   Chin, S C.1989, The Sustainability of Shifting Cultivation, World Rainforest Movement, Penang

9.   de Beer, J. and McDermott, M. 1989, The Economic Value of Non-timber Forest Products in Southeast Asia, Netherlands Committee for IUCN, Amsterdam

10. Shiva, V. 1989, *Staying Alive*, Zed Books, London, p59

11. Randhawa, M S. 1989, A History of Agriculture in India, Indian Council of Agricultural Research, p97

12. Panday, K K. 1982, Fodder Trees and Tree Fodder in Nepal, Swiss Development Cooperation, Berne

13. Singh, S P. and Berry, A. 1985, Forestry Land Evaluation at District Level, FAO Bangkok

14. Mahat, T B S. 1987, Forestry - Farming Linkages, in the Mountains, ICIMOD, Kathmandu

15. WRM, 1990, *The Battle for Sarawak's Forests*, World Rainforest and SAM publication, Penang

16 Schlick, S. 1920, *Systems of Silviculture*

17. Troup, R S. 1916, *Silviculture Systems*, Oxford University Press, Oxford

18. Bethel, J. 1984, Sometimes the Word is 'Weed', *Forest Management*, June 1984, pp17-22

19. Shiva, V. Bandyopadhyay, J. and Sharatchandra, H C. 1981 *The Social Ecological and Economic Impact of Social Forestry in Kolar*, IIM, Bangalore

20. Quereshi, T M. 1967, The Concept of Fast Growth in Forestry and the Place of Indigenous Fast Growing Broad Leaved species, *Proceedings of the Eleventh Silvicultural Conference*, FRI, Dehra Dun

21. Chaturvedi, A N. 1983, Eucalyptus for Farming, U P Forest Department, Lucknow

22. Patel, V J. 1984, Rational Approach Towards Fuelwood Crisis in Rural India, Jivarajbhai Patel Agroforestry Centre, Surendrabag - Kardej p10

23. Gupta, R K. Aggarwal, M C. and Hira Lal, 1984, Correlation Studies of Phytomass of Fodder Trees with Growth Parameters, *Soil Conservation Bulletin*, Dehra Dun, p9

24. Singh, R V.1982, Fodder Trees of India, Oxford, New Delhi

25. Lappe, F. and Collins, J. 1982, *Food First*, Abacus, p114

26. Yegna Iyengar, A K. 1944, Field Crops of India, BAPPCO, Bangalore p30

27.   Swaminathan, M S. 1983, Science and the Conquest of Hunger
      Concept, Delhi, p113

28.   Shah, C H. (ed) Agricultural Development of India, Orient
      Longman, Delhi 1979, p xxxii

29.   Richaria, R H. 1986, Paper presented at Seminar on Crisis in Modern
      Science, Penang

30.   Yegna Iyengar, op cit, p30

31.   Wright Angus, 1984, Innocents Abroad: American Agricultural
      Research in Mexico, in Jackson, W. et al (ed) *Meeting the Expectations
      of the Land*, North Point Press, San Francisco, 1984

32.   Schlich, op cit

33.   Caufield, op cit p177

34.   UNESCO 1985, Tropical Forest Ecosystems, Paris

35.   Caufield, op cit, p178

36.   FAO, 1986, Tropical Forest Management, Rome

37.   Fahser, L.   1986, The Ecological Orientation of the Forest Economy,
      lecture given at Faculty of Forestry Science, University of Freiburg
      im Breisgan

38.   Shiva, V. and Bandyopadhyay, J. 1985, Ecological Audit of Eucalyp-
      tus Cultivation, Research Foundation, Dehra Dun

39.   Shiva, V. 1989, The Violence of the Green Revolution, Research
      Foundation of Science and Ecology, Dehra Dun

2 Biodiversity: A Third World Perspective

The Crisis of Diversity

DIVERSITY IS THE CHARACTERISTIC of nature and the basis of ecological stability. Diverse ecosystems give rise to diverse life forms, and to diverse cultures. The co-evolution of cultures, life forms and habitats has conserved the biological diversity on this planet. Cultural diversity and biological diversity go hand in hand.

Communities everywhere in the world have developed knowledge and found ways to derive livelihoods from the bounties of nature's diversity, in wild and domesticated forms. Hunting and gathering communities use thousands of plants and animals for food, medicine and shelter. Pastoral, peasant and fishing communities have also evolved knowledge and skills to derive sustainable livelihoods from the living diversity on the land and in rivers, lakes and seas. The deep and sophisticated ecological knowledge of biodiversity has given rise to cultural rules for conservation reflected in notions of sacredness and taboos.

Today, however, the diversity of ecosystems, life forms and ways of life of different communities is under threat of extinction. Habitats have been enclosed or destroyed, diversity has been eroded and livelihoods deriving from biodiversity are threatened.

Tropical moist forests cover only 7% of the earth's land surface but contain at least half of the earth's species. Deforestation in these regions is continuing at a rapid pace, with very conservative estimates suggesting rates as high as 6.5% in Cote d'Ivoire and averaging about 0.6% per year (about 7.3 million ha) for all tropical countries. At this rate, which is a net figure, and incorporating reforestation and natural growth, all closed tropical forests would be cleared within 177 years (FAO, 1981). Raven (1988) estimates that about 48% of the world's plant species occur in or around forest areas where over more than 90% of their area will be destroyed during the next 20 years, leading to about a quarter of those species being lost. Wilson (1988) has estimated that the current extinction rate is 1000 species a year. By the 1990s, the figure is expected to rise to ten thousand species a year (one species an hour). During the next 30 years, one million species could be erased.

Biological diversity in marine ecosystems is also remarkable, and coral reefs are sometimes compared with tropical forests in terms of diversity (Connell, 1978). Marine habitats and marine life are under severe threat; with the destruction of diversity, the fisheries base in most coastal regions of the world is on the verge of collapse.

The erosion of diversity is also very severe in agricultural ecosystems. Crop varieties have disappeared, and cultivation during the 'Green Revolution' phase shifted from hundreds and thousands of crops to wheat and rice derived from a very narrow genetic base. The wheat seeds that spread worldwide from the International Centre for Maize and Wheat Improvement (CIMMYT) through Norman Borlaug and his 'wheat apostles' were the result of nine years of experimenting with Japanese 'Norin' wheat. 'Norin', released in Japan in 1935, was a cross between Japanese dwarf wheat called 'Daruma' and American wheat called 'Faltz' which the Japanese government had imported from the US in 1887. The 'Norin' wheat was brought to the US in 1946 by Dr D C Salmon, an agriculturist acting as a US military adviser in Japan, and further crossed with American seeds of the variety called 'Bevor' by US Department of Agriculture scientist Dr Orville Vogel. Vogel in turn sent it to Mexico in the 1950s where it was used by Borlaug, who was on the Rockefeller Foundation staff, to develop his well-known Mexican varieties. Of the thousands of dwarf seeds created by Borlaug, only three were used to create the 'Green Revolution' wheat plants which were spread worldwide. On this narrow and alien genetic base, are the food supplies of millions precariously perched.[1]

Over the last half century, India has probably grown over 30,000 different indigenous varieties or land races of rice. The situation has altered drastically in the past 15 years, however, and Dr H K Jain, Director of the Indian Agricultural Research Institute in New Delhi predicts that in another 15 years this enormous rice diversity will be reduced to no more than 50 varieties, with the top ten accounting for over three-quarters of the subcontinent's rice acreage.[2]

Livestock populations are also being homogenised and their diversity is being irreversibly lost. The carefully evolved pure breeds of cattle in India are on their way to extinction. The Sahiwal, Red Sindhi, Rathi, Tharparkar, Hariana, Ongole, Kankreji, Gir are cattle breeds developed for the different eco-niches where they had to survive and support the needs of local communities. Today they are being systematically substituted by cross breeds of Jersey and Holstein Cows.

With animals disappearing as an essential component of farming systems, and their contribution of organic fertility being substituted by chemical fertilisers, soil, fauna and flora have also gone extinct. The locally specific nitrogen-fixing bacteria, fungi that facilitate nutrient intake through mycorrhizal association, predators of pests, pollinators and seed dispersers, and other species that co-evolved over centuries to provide environmental services to traditional agrosystems have become extinct, or have had their genetic base dramatically narrowed. Deprived of the flora with which they co-evolved, soil microbes also disappear (Norgaard, 1988).

Biodiversity erosion starts a chain reaction. The disappearance of a species is related to the extinction of innumerable other species with which it is inter-related through food webs and food chains, and about which humanity is totally ignorant. The crisis of biodiversity is not just a crisis of the disappearance of species which have the potential of spinning dollars for corporate enterprises by serving as industrial raw material. It is, more basically, a crisis that threatens the life-support systems and livelihoods of millions of people in Third World countries.

Primary Threats to Biodiversity

(i) *Primary causes*

There are two primary causes for the large scale destruction of biodiversity:-

The first is habitat destruction due to internationally financed mega-projects such as the building of dams and highways, and mining operations in forested regions rich in biological diversity.

The second primary cause for the destruction of biodiversity in areas under cultivation is the technological and economic push to replace diversity with homogeneity in forestry, agriculture, fisheries and animal husbandry. The Green Revolution in agriculture, the White Revolution in dairying and the Blue Revolution in fisheries are revolutions based on the deliberate replacement of biological diversity with biological uniformity and monocultures.

a)   *Biodiversity destruction due to development projects in forest areas*

The Narmada dams will submerge large areas of forests in the Narmada Valley in India. The Sardar Sarovar project will submerge 11,000 ha and the Narmada Sagar will submerge nearly 40,000 ha of forest land. Besides direct destruction of biodiversity in these forests, the submergence will irreversibly destroy the survival base of tribals in the region.

In Thailand, the Nam Choan Dam would have flooded the valley land of the Tung Yai and Huai Kha Khaeng wildlife sanctuaries, which together comprise the largest intact block of forested land set aside for wildlife conservation in Thailand. The dam thus threatened to destroy the habitat of the largest remaining populations of elephant and bantcug, as well as a variety of other threatened or endangered species such as tiger, gaur, and tapir, and birds like the green pea fowl. (Tuntawiroon and Samotsa-Korn 1984).

In Brazil, the Grande Carajas Programme involving the Tucurui dams, iron ore and bauxite mining and processing industry threatens biodiversity and cultural diversity in the Amazon. Amazonia harbours more wildlife than any other place on Earth, both per unit area and as a subcontinental region. There are estimated to be 'over 50,000 species of higher plants, at least an equal number of fungi, a fifth of all the birds on our planet, at least 3,000 species of fish, amounting to ten times the number of fish in all the rivers of Europe, and insect species numbering in the uncounted millions' in the Amazonia.

The great age and size of the forests, their favourable climate (hot and moist), the fact that they have remained undisturbed for millenia, and the presence of very high concentrations of species in particular areas (known as *Peistocene refugia)* have all contributed to the region's unparalleled diversity. For instance, a typical hectare of Amazonian forest contains between 200-300 different varieties of trees alone.[3]

During the filling of the Tucurui reservoir which flooded at least 2,150 square kilometres of rainforest over many months, an attempt was made to rescue drowning animals. In one ten-day period, 4,037 mammals, 4,848 reptiles, 6,293 insects such as giant scorpions and spiders, 717 birds and 30 amphibians were captured by men in boats – some 15,925 creatures from one part of the lagoon. Brazilian ecologists estimated that this total was a tiny fraction of the actual number held by the forest.

The 10% of the world's species that occur in Amazonia are not uniformly spread, but cluster throughout the river basin. Most are endemic or have limited distributions. Inevitably, high diversity means that there are relatively few individuals of any one species. The more development intensifies, the greater the likelihood of extinctions. In regions like Carajas, where single projects involve the clearance of thousands of square kilometres of forests, not just individual species but whole habitats are rapidly disappearing.[4]

(b) *Displacement of biodiversity by monocultures*

According to the dominant paradigm of production, diversity goes against productivity, which creates an imperative for uniformity and monocultures. This has generated the paradoxical situation in which plant improvement has been based on the destruction of the biodiversity which it uses as raw material. The irony of plant and animal breeding is that it destroys the very building blocks on which the technology depends. Forestry development schemes introduce monocultures of industrial species like eucalyptus and push into extinction the diversity of local species which fulfil local needs. Agricultural modernization schemes introduce new and uniform crops into farmers' fields and destroy the diversity of local varieties. In the words of Professor Garrison Wilkes of the University of Massachusetts, this is analogous to taking stones from a building's foundation to repair the roof. This strategy of basing productivity increase on the destruction of diversity is dangerous and unnecessary.

Not until diversity is made the logic of production can diversity be conserved. 'Improvement' from the corporate viewpoint, or from the viewpoint of western agricultural or forestry research, is often a loss for the Third World, especially the poor in the Third World. There is therefore no inevitability that production should act against diversity. Uniformity as a pattern of production becomes inevitable only in a context of control and profitability.

The spread of monocultures of 'fast-growing' species in forestry and 'high-yielding varieties' in agriculture has been justified on grounds of increased productivity. All technological transformation of biodiversity is justified in the name of 'improvement' and increased 'economic value'. However, 'improvement' and 'value' are not neutral terms. They are contextual and value-laden. Improvement of tree species means one thing for a paper corporation which needs pulping wood and an entirely different thing for a peasant who needs fodder and green manure. Improvement of crop species means one thing for a food processing industry and something totally different for a self-provisioning farmer.

The categories of 'yield', 'productivity' and 'improvement' which have emerged from the corporate viewpoint have, however, been treated as universal and value-neutral. Thus, all tree planting programmes financed by international institutions in recent years and encouraged by the Tropical Forestry Action Plan (TFAP) have spread fast-growing eucalyptus monocultures across Asia, Africa and Latin America. The only fast growth to which eucalyptus contributes is pulp wood – it is not fast growing in terms of yield of wood for other purposes, and in terms of yields of non-woody biomass for fodder it has zero yields since its leaves are not eaten by cattle. Given that the industrial sector does not benefit from the diversity of species and uses of trees, forestry programmes deliberately destroy diversity in order to increase yields of industrial raw material.

Viewing diversity as weeds leads to the extinction of that diversity which has high ecological and social value even when it does not profit industry. The pattern of destruction of diversity has been the same in both forestry and agriculture.

Plant improvement in agriculture has been based on the enhancement of the yield of a desired product at the expense of unwanted plant parts. The 'desired' product is however not the same for agri-business and a Third World peasant. Which parts of a farming system will be treated as 'unwanted' depends on one's class and gender. What is unwanted for agribusiness may be wanted by the poor, and when it squeezes out those aspects of biodiversity, agriculture 'development' fosters poverty and eco-logical decline.

In India, the 'high-yielding' strategy of the Green Revolution squeezed out pulses and oilseeds which were essential for nutri-tion and soil fertility. The monocultures of the dwarf varieties of wheat and rice also squeezed out the straw which was essential for fodder and fertilizing the soil. The yields were 'high' for the purposes of centralised control of the food-grain trade, but not in the context of diversity of species and products at the level of the farm and the farmer. Productivity therefore differs depending on whether it is measured in a framework of diversity or uniformity.

(ii) *Secondary Causes of Biodiversity Erosion*

The dominant view ignores the primary causes of biodiversity destruction and instead focuses on secondary causes such as population pressure. However, stable communities, in harmony with their ecosystem, always protect biodiversity. It is only when populations are displaced by dams, mines, factories, and commercial agriculture that their relationship to biodiversity becomes antagonistic rather than co-operative. The displacement of people and displacement of diversity goes hand in hand, and displaced people further destroying biodiversity is a second order effect of the primary causes of destruction identified above.

Effects of Biodiversity Erosion

The erosion of biodiversity has serious ecological and social consequences since diversity is the basis of ecological and social stability. Social and material systems devoid of diversity are vulnerable to collapse and breakdown.

i) *Ecological vulnerability of monocultures of 'improved varieties'*

Case A: In 1970-71, America's vast cornbelt was attacked by a mysterious disease, later identified as 'race T' of the fungus *Helminisporium maydis* which caused the Southern Corn Leaf Blight as the epidemic was called. It left ravaged corn fields with withered plants, broken stalks and malformed or completely rotten cobs with a grayish powder. The strength and speed of the Blight was a result of the uniformity of hybrid corn, most of which had been derived from a single Texas male sterile line. The genetic make-up of the new hybrid corn which was responsible for its rapid and large scale breeding by seed companies was also responsible for its vulnerability to disease. At least 80% of the hybrid corn in America in 1970 contained the Texas male sterile cytoplasm. As a University of Iowa pathologist wrote, 'Such an extensive, homogeneous acreage is like a tinder-dry prairie waiting for a spark to ignite it.'

A National Academy of Sciences study *'Genetic Vulnerability of Major Crops'* stated: 'The corn crop fell victim to the epidemic because of a quirk in the technology that had redesigned the corn plants of America until in one sense, they had become as alike as identical twins. Whatever made one plant susceptible made them all susceptible.' (Doyle, 1988)

Case B: In 1966, the International Rice Research Institute released a 'miracle' rice variety – IR-8, which was quickly adopted for use through Asia. IR-8 was particularly susceptible to a wide range of disease and pests: in 1968 and 1969 it was hit hard by bacterial blight and in 1970 and 1971 it was ravaged by another tropical disease called tungro. In 1975, Indonesian farmers lost half a million acres of Green Revolution rice varieties to leaf hoppers. In 1977, IR-36 was developed to be resistant to 8 major diseases and pests including bacterial blight and tungro. However this was attacked by two new viruses called 'ragged stunt' and 'wilted stunt'.

The vulnerability of rice to new pests and disease due to monocropping and a narrow genetic base is very high. IR-8 is an advanced rice variety that came from a cross between an Indonesian variety called 'Pea' and another from Taiwan called 'Dee-Geo-Woo-Gen'. IR-8, Taichung Native 1 (TN1) and other varieties were brought to India and became the basis of the All India Coordinated Rice Improvement Project to evolve dwarf, photoinsensitive, short duration, high yielding varieties of rice suited to high fertility conditions. The large scale spread of exotic strains of rice with a narrow genetic base was known to carry the risk of the large-scale spread of disease and pests. As summarised in a publication titled 'Rice Research in India – An Overview' from CRRI,

The introduction of high yielding varieties has brought about a marked change in the status of insect pests like gall midge, brown

planthopper, leaf folder, whore maggot etc. Most of the high yielding varieties released so far are susceptible to major pests with a crop loss of 30 to 100%... Most of the HYVs are the derivatives of TN1 or IR-8 and therefore, have the dwarfing gene of Dee-Geo-Woo-Gen. The narrow genetic base has created alarming uniformity, causing vulnerability to disease and pests. Most of the released varieties are not suitable for typical uplands and low-lands which together constitute about 75% of the total rice area of the country.[5]

The 'miracle' varieties displaced the diversity of traditionally grown crops, and through the erosion of diversity, the new seeds became a mechanism for introducing and fostering pests. Indigenous varieties or land races are resistant to locally occuring pests and diseases. Even if certain diseases occur, some of the strains may be susceptible, while others will have the resistance to survive. Crop rotations also help in pest control. Since many pests are specific to particular plants, planting crops in different seasons and different years causes large reductions in pest populations. On the other hand, planting the same crop over large areas year after year encourages pest build ups. Cropping systems based on diversity thus have built-in protection.

ii) *Social vulnerability of homogeneous systems*

The two principles on which the production and maintenance of life is based are:

(a) the principle of diversity, and
(b) the principle of symbiosis and reciprocity, often also called the law of return.

The two principles are not independent but interrelated. Diversity gives rise to the ecological space for give and take, for mutuality and reciprocity. Destruction of diversity is linked to the creation of monocultures, and with creation of monocultures, the

Monocultures are associated with external
inputs, centralised regulation and high
vulnerability to ecological breakdown.

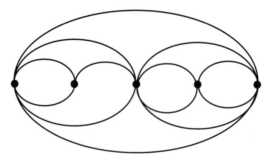

Systems based on diversity are associated with
decentred self-regulation and high resilience.

self-regulated and decentralised organisation of diverse systems gives way to external inputs and external and centralised control. Schematically the transformation can be imaged figures as illustrated on page 76.

Sustainability and diversity are ecologically linked because diversity offers the multiplicity of interactions which can heal ecological disturbance to any part of the system. Nonsustainability and uniformity means that a disturbance to one part is translated into a disturbance to all other parts. Instead of being contained, ecological destabilisation tends to be amplified. Closely linked to the issue of diversity and uniformity is the issue of productivity. Higher yields and higher production have been the main push for the introduction of uniformity and the logic of the assembly line. The imperative of growth generates the imperative for monocultures. Yet this growth is, in large measure, a socially-constructed, value-laden category. It exists as a 'fact' by excluding and erasing the facts of diversity and production through diversity.

Diverse systems have multiple outputs and yields, and much of these outputs flow back within the system to allow for 'low-external-input' production, so that production is possible without access to purchasing power, credits and capital. Livestock and crops help maintain each other's productivity symbiotically and sustainably. Different crop varieties also maintain each other eg. corn and beans, millets and pulses, where the legume provides nitrogen for the main cereal crop through nitrogen fixation.

In addition to providing ecological stability, diversity also ensures diverse livelihoods and provides for multiple needs through reciprocal arrangements.

Homogeneous and one dimensional production systems break up community structure, displace people from diverse occupations, and make production dependent on external inputs and external markets. This generates political and economic vulnerability and instability because the production base is ecologically unstable and commodity markets are economically unstable.

Negros in the Philippines is an economic disaster because its entire economy depended on sugarcane, and when sugar substitutes were derived from corn, there was no longer a market for sugarcane. The vulnerability of Africa is extremely high because colonialism introduced exclusive dependence on monocultures of cash crops for exports and displacement of biodiversity for local food needs. Many African countries rely on single crops for export earnings.

With the emergence of the new biotechnologies and the industrial production of substitutes for the biological products from plantation crops, severe dislocation of the economy and society in these countries can be expected.

First World Bio-imperialism and North-South Conflicts

The wealth of Europe in the colonial era was to a large extent, based on the transfer of biological resources from the colonies to the centres of imperial power, and the displacement of local biodiversity in the colonies by monocultures of raw material for European industry.

A W Crosby has called the biological transfer of wealth from the Americas to Europe the 'Columbian exchange', because with Columbus' arrival in America started the mass transfer of maize, potatoes, squash, tomatoes, peanuts, common beans, sunflowers and other crops across the Atlantic.

Various spices, sugar, bananas, coffee, tea, rubber, indigo, cotton and other industrial crops began to make their move to new production sites under the control of newly emerging colonial powers and their state backed trading companies.

Violence and control were an intrinsic part of this process by which the North accumulated capital and wealth by gaining control over the biological resources of the South. Destroying the biodiversity it could use or control was the other less visible side of this process of colonisation.

In 1876 the British smuggled rubber out of Brazil and introduced it in its colonies in Sri Lanka and Malaya. The Brazilian rubber industry collapsed and famine replaced the rubber business.

The Dutch cut down 75% of the clove and nutmeg stands in the Moluccas and concentrated production on three heavily guarded islands.

Physical violence might no longer be the main instrument of control, but control of the Third World's biodiversity for profits is still the primary logic of North-South relationships on biodiversity. The large scale introduction of monocultures in the Third World through the Green Revolution was spearheaded by the International Centre for Wheat and Maize Improvement (CIMMYT) in Mexico and International Rice Research Institute (IRRI) in the Philippines, controlled by the Consultative Group on International Agricultural Research (CGIAR), which was launched by the World Bank in 1970.

In the Philippines, IRRI seeds acquired the name 'seeds of imperialism'. Robert Onate, President of the Philippines Agricultural Economics and Development Association observed that IRRI practices had created a new dependence on agrochemicals, seeds and debt. 'This is the Green Revolution Connection,' he remarked, 'New seeds from the CGIAR global crop seed systems which will depend on the fertilizers, agrichemicals and machineries produced by conglomerates of the Transnational Corporations.'

The International Bureau for Plant Genetic Resources (IBPGR) which is run by the CGIAR system was specifically created for the collection and conservation of genetic resources. However, it has emerged as an instrument for the transfer of resources from the South to the North. While most genetic diversity lies in the South, of the 127 base collections of IBPGR, 81 are in the industrialised countries, and 29 are in the CGIAR system which is controlled by the governments and corporations of the industrialised countries in the North. Only 17 are in the national collections of Third World countries. Of the 81 base collections in the North, 10 are in the hands of the countries that fund IBPGR.

The US has accused countries of the Third World as engaging in 'unfair trading practice' if they fail to adopt US patent laws which allow monopoly rights in life forms. Yet it is the US which has engaged in unfair practices related to the use of Third World genetic resources. It has freely taken the biological diversity of the Third World to spin millions of dollars of profits, none of which have been shared with Third World countries, the original owners of the germ plasm.

According to Prescott-Allen, wild varieties contributed US$340 million per year between 1976 and 1980 to the US farm economy. The total contribution of wild germ plasm to the American economy has been US$66 billion, which is more than the total international debt of Mexico and the Philippines combined. This wild material is 'owned' by sovereign states and by local people.

A wild tomato variety (*Lycopresicon chomrelewskii*) taken from Peru in 1962 has contributed US$8 million a year to the American tomato processing industry by increasing the content of soluble solids. Yet none of these profits or benefits have been shared with Peru, the original source of the genetic material.

(i) *Drug firms rob Third World's medicinal plants*

The pharmaceutical industry of the North has similarly benefited from free collection of tropical biodiversity. The value of the South's germplasm for pharmaceutical industry ranges from an estimated US$4.7 billion now to US$47 billion by the year 2000.

As drug companies realise that nature holds rich sources of profit they begin to covet the potential wealth of tropical moist forests as a source for medicines. For instance, the periwinkle plant from Madagascar is the source of at least 60 alkaloids which can treat childhood leukaemia and Hodgkin's Disease. Drugs derived from this plant bring in about US$160 million worth of sales each year. Yet another plant, *Rauwolfa serpentina*, from India is the base for drugs which sell up to US$260 million a year in the US alone.

Unfortunately, it has been estimated that with the present rate of destruction of tropical forests, 20-25% of the world's plant species will be lost by the year 2000. Consequently, major pharmaceutical companies are now screening and collecting natural plants through contracted third parties. For instance, a British company, Biotics, is a commercial broker known for supplying exotic plants for pharmaceutical screening by inadequately compensating the Third World countries of origin. The company's officials have actually admitted that many drug companies prefer 'sneaking plants' out of the Third World than going through legitimate negotiating channels.

Screening and collection cover plants, bacteria, algae, fungi, protozoa and a wide range of marine organisms including corals, sponges and anemones.

Another method is that of the US National Cancer Institute which has sponsored the single largest tropical plant collecting efforts by recruiting the assistance of ethno-botanists, who in turn siphon off the traditional knowledge of indigenous peoples without any compensation.

(ii) *First World bio-imperialism*

In spite of the immeasurable contribution that Third World biodiversity has made to the wealth of industrialised countries, corporations, governments and aid agencies of the North continue to create legal and political frameworks to make the Third World pay for what it originally gave. The emerging trends in global trade and technology work inherently against justice and ecological sustainability. They threaten to create a new era of bio-imperialism, built on the biological impoverishment of the Third World and the biosphere.

The intensity of this assault against Third World genetic resources can be seen from the pressure exerted by major drug and agricultural input companies and their home governments on international institutions such as the General Agreement on Tariffs and Trade (GATT) and the FAO to recognise such resources as a 'universal heritage' in order to guarantee them free access to the raw materials. International patent and licensing agreements will increasingly be used to secure a monopoly over valuable genetic materials which can be developed into drugs, food, and energy sources.

Limitations of the Dominant Approaches to Biodiversity Conservation

The dominant approaches to biodiversity conservation suffer from the limitations of a northern bias, and a blindness to the role of the North in the destruction of biodiversity in the South.

'Conserving the World's Biological Diversity' (a study released by the World Bank, the World Resources Institute, the International Union for the Conservation of Nature Resources and the World wide Fund for Nature) has undoubtedly emerged from the North. However, even this report suffers from a biased analysis and biased prescriptions.

(i) *Neglect of primary causes of destruction*

In this report, while the crisis of erosion is focused on as an exclusively tropical, and Third World phenomenon, the thinking and planning of biodiversity conservation is projected as a monopoly of institutes and agencies based in and controlled by the industrial North. It is as if the mind and the solutions are in the North, while the matter and the problems are in the South. This polarity and dualism underlies the basic shortcomings of the book, which could more honestly have been titled 'The North Conserving the South's Biological Diversity'.

It is of course true that the tropics are the cradle of the planet's biological diversity, with an incomparable multiplicity and variability of ecosystems and species. However, not only is erosion of diversity as great a crisis in the North, it is also in the North that the roots of the South's crisis of diversity lie. These aspects of the destruction of diversity are not addressed in the book.

Closely related to the book's neglect of forces and factors in the North as part of the problem is its neglect of the crisis of diversity in what are viewed as 'production' spheres – forestry, livestock and agriculture. Among the causes identified as leading to the loss of biological resources are forest clearing and burning, overharvesting of plants and animals, and indiscriminate use of pesticides. In the past 20-30 years, however, in addition to these factors, there has been a deliberate substitution of diversity by uniformity of crops, trees and livestock – through development projects financed by aid from international agencies.

The report thus ignores the two primary causes of biodiversity destruction which are global in character,and focuses on secondary and minor causes which are often local in character. It therefore blames the victims of biodiversity destruction for the destruction, and places responsibility for conservation in the hands of the sources of destruction.

(ii) *Disease offered as cure*

The World Bank, which continues to introduce biodiversity action plans, has for the past 10 years been financing the destruction of genetic diversity in the Third World. It financed the Green Revolution which replaced genetically diverse indigenous cropping systems in the Third World with vulnerable, genetically uniform monocultures. It contributed to genetic erosion through the encouragement of centralised research institutions controlled by the Consultative Group on International Agriculture Research (CGIAR), which it launched in 1970.

The Tropical Forest Action Plan (TFAP) which is cited as an example of a strategy for conserving habitats has been responsible for the destruction of biodiversity in both natural forests and agricultural ecosystems. Large scale introduction of monocultures of eucalyptus and other industrial species has been accelerated under the TFAP, displacing indigenous tree, crop and animal species. In effect the TFAP has become an instrument for giving public subsidies to multinational corporations such as Shell and Jaako Poyry in Asia and Latin America.

(iii) *Who produces, who consumes biodiversity?*

The northern bias of the World Bank/IUCN/WRI/WWF report is also evident in its analysis of the value of biodiversity. In the self provisioning economies of the Third World, producers are simultaneously consumers and conservers. In fact, it is recognised that 'the total genetic change achieved by farmers over the millenia was far greater than that achieved by the hundred or two years of more systematic science based efforts'. (Kloppenberg, 1988).

If this contribution to knowledge and development of biodiversity is recognised, farmers and tribals are the original producers, and corporate and public sector scientists consume their finished products as raw material for commodities. The dominant approach puts this relationship of producer and consumer on its head.

Probably the authors' treatment of Northern agencies as part of the solution rather than part of the problem is related to their economistic approach. In the chapter on 'Values of biological diversity', it is recognised that biological resources have social, ethical, cultural and economic values. 'But', the authors proceed to say:

...in order to compete for the attention of government decision makers in today's world, policies regarding biological diversity first need to demonstrate in economic terms the value of biological resources to a country's social and economic development.

The economic values of biological resources are then divided into the following categories:

● 'consumptive value' – value of products consumed directly without passing through a market, such as firewood, fodder and game meat;

● 'productive use value' – value of products commercially exploited; and

● 'non-consumptive use value' – indirect value of ecosystem functions, such as watershed protection, photo-synthesis, regulation of climate and production of soil.

An interesting value framework has thus been constructed which predetermines analysis and options. If the Third World poor, who derive their livelihoods directly from nature, only 'consume', and the trading and commercial interests are the only 'producers', it follows quite naturally that the Third World is responsible for the destruction of its biological wealth, and the North alone has the capacity to conserve it. This ideologically constructed divide between consumption, production and conservation hides the political economy of the processes which underlie the destruction of biological diversity.

Defining production as consumption and consumption as production also matches the demand for intellectual property rights of the North, and denies the intellectual contributions of those in the South who are the primary producers of value.

(iv) *Commercialised conservation*

The economistic bias narrows down conservation options to a commercialised approach in which both the means and ends of conservation are financial values on the market.

Commercialised conservation is linked to the emergence of new biotechnologies which have transformed the genetic resources of this planet into the raw material for industrial production of food, pharmaceuticals, fibres, energy, etc. Commercialised conservation measures and justifies the value of conservation in terms of its present or future use for profits. It does not take into account that this will wipe out genetic diversity. Biodiversity conservation here is seen only in terms of setting aside reserves in undisturbed ecosystems for the purpose of conservation. This schizophrenic approach to biodiversity, which adopts a policy of destruction of diversity in production processes and a policy of preservation in 'set-asides', cannot be effective in the conservation of species diversity. Biodiversity cannot be conserved unless production itself is based on a policy of preserving diversity.

Exclusive dependence on economic value as the reason for conservation is the wrong place to initiate a conservation programme. As Ehrenfeld has noted: 'By assigning value to diversity we merely legitimise the process that is wiping it out, the process that says, "The first thing that matters in any important decision is the tangible magnitude of the dollar costs and benefits...." If conservation is to succeed, the public must come to understand the inherent wrongness of the destruction of biological diversity.' (Ehrenfeld, 1988).

(v) *The reductionist approach*

The dominant approach to biodiversity is inadequate for conservation both because it values biodiversity only as a commodity, but also because it perceives biodiversity in a fragmented and atomised form. It views biodiversity merely as an arithmetic, numerical, additive category. Thus 'conserving the world's biological diversity' uses biodiversity as an 'umbrella term for the degree of nature's variety, including both the number and frequency of ecosystems, species or genes in a given assemblage' (McNeely et.al., 1990). This leads to a reductionist approach to conservation, which serves commercial objectives well, but fails to fulfil ecological criteria.

Ex situ conservation in high-tech gene banks is the dominant response to conservation of biodiversity. This approach is both static and centralised. It is an efficient means of conservation of raw material in the form of germ plasm collection. However, it has its limitations both because it removes control over biodiversity from local communities from whose custody the germ plasm has been taken away, and it removes biodiversity from the habitats where the diversity would evolve and adapt under changing environmental conditions.

From Bio-imperialism to Bio-democracy

Conserving Biodiversity on the basis of ecology and equity

(i) *Ecology, equity and efficiency*

An ecologically sustainable and just approach to biodiversity conservation needs to begin by halting and reversing the primary threats to biodiversity. This involves stopping aid and incentives for the large scale destruction of habitats where biodiversity thrives, and stopping subsidies and public support for displacement of diversity by centralised and homogeneous systems of production in forestry, agriculture, fisheries and animal husbandry. Since the drive for this destruction comes from international aid and financing, the beginning for stopping biodiversity destruction and for starting conservation has to be made at that level. In parallel, support needs to be given to ways of life and systems of production that are based on the conservation of diversity, and which have been marginalised by the dominant pattern of development.

Ecologically, this shift involves recognition of the value of diversity in itself. As Ehrenfeld has stated: 'Value is an intrinsic part of diversity.' All life forms have an inherent right to life, and that should be the overriding reason for not allowing species extinction to take place.

At the social level, the values of biodiversity in different cultural contexts need to be recognised. Sacred groves, sacred seeds, sacred species have been cultural means for treating biodiversity as inviolable, and present us with the best examples of conservation. In addition, we need to recognise that market value and dollar value is only a limited value which is often perverse for biodiversity. There are other values of biodiversity, such as those of providing meaning and sustenance, and these values need not be treated as subservient and secondary to market values.

The recognition of community rights to biodiversity, and farmers' and tribals' contributions to the evolution and protection of biodiversity also need to be recognised – by treating their knowledge systems as futuristic, not as primitive.

At the economic level, if biodiversity conservation is to be aimed at conserving life, rather than profits, then the incentives given to biodiversity destruction and the penalties that have become associated with biodiversity conservation need to be removed. If a biodiversity framework guides economic thinking rather than the other way around, it becomes evident that the so-called high production of homogeneous and uniform systems is an artificial measure, which is artificially maintained through public subsidies. If half a calorie of energy produces one calorie of food in non-industrial biodiversity based systems, and 10 calories of energy produce one calorie of food in a homogeneous industrial system, it is clearly not efficiency and productivity that pushes the displacement of the former by the latter. Productivity and efficiency need to be redefined, reflecting the multiple input, multiple output and internal input systems characterised by biodiversity.

In addition, the perverse logic of financing biodiversity conservation by a small percentage of profits generated by biodiversity destruction amounts to giving licence to destruction, and reduces conservation into an exhibit, not a basis of living and producing. The disadvantages for conserving systems arise from privileges given to destroying systems, and conservation cannot be achieved by extending those privileges and deepening the disadvantages. Third World governments need to remember that one cannot protect one's house against theft by begging the thief to give back a small share of the loot. Protection comes from not allowing theft to take place in the first place.

Ecology, equity and efficiency meet in biodiversity, while they are in opposition with each other in monocultures and homogeneous systems. Diversity ensures ecological stability. Diversity ensures multiple livelihoods and social justice. Diversity also ensures efficiency in a multidimensional context. On the other hand, uniformity creates:
- (a) ecological instability;
- (b) external control, which leads to displacement of livelihoods;
- (c) efficiency in a one dimensional framework, but undermines it at the systems level.

(ii) *Who controls biodiversity?*

Neither ecological sustainability nor livelihood sustainability can be ensured without a just resolution of the issue of who controls biodiversity.

Until recent times, it was local communities who have used, developed and conserved biological diversity, who have been custodians of the biological wealth of this planet. It is their control, their knowledge and their rights that need to be strengthened if the foundations of biodiversity conservation are to be strong and deep. This strengthening has to be done through local action, national action and global action.

After centuries of the gene-rich South having contributed biological resources freely to the North, Third World governments are no longer willing to have biological wealth taken for free and sold back at exorbitant prices to the Third World as 'improved' seeds and packaged drugs. From the Third World viewpoint, it is considered highly unjust that the South's biodiversity be treated as the 'common heritage of mankind', and the return flow of biological commodities be patented, priced and treated as private property of northern corporations.

This new inequality and injustice is being forced on the Third World through the patent system and intellectual property rights by GATT, the World Bank and the US Trade Act. The new North-South asymmetries it will generate make for an unstable world and are of course an issue of major concern. Equally serious is the undermining of the sovereignty of the Third World.

But much more serious is the total erosion of sovereignty of local communities, the original custodians of biodiversity, and the sovereignty of the diversity of life-forms which are our partners in co-evolution, not merely mines of genes to be exploited at will for profits and control.

Putting value on the gene through patents makes biology stand on its head. Complex organisms which have evolved over millenia in nature, and through the contributions of Third World peasants, tribals and healers are reduced to their parts, and treated as mere inputs into genetic engineering. Patenting of genes thus leads to a devaluation of life-forms by reducing them to their constituents and allowing them to be repeatedly owned as private property. This reductionism and fragmentation might be conven-ient for commercial concerns, but it violates the integrity of life as well as the common property rights of Third World peoples. On these false notions of genetic resources and their ownership through intellectual property rights are based the 'bio-battles' at FAO and the trade wars at GATT.

To redress the North-South imbalance and to recognise the contributions of local communities to the development of biodiversity, it is imperative that the regime based on bio-imperialism be replaced by structures based on biodemocracy. Gandhi has shown us that absolute power based on unethical and un-democratic foundations can only be challenged by a resurgence of the ethical and democratic.

Biodemocracy involves the recognition of the intrinsic value of all life forms and their inherent right to exit. It also involves the recognition of original contributions and rights of communities which have co-evolved with local biodiversity.

Biodemocracy entails that nation states protect these prior rights from erosion by corporate claims to private property in life forms through patents and intellectual property rights.

The deeper the devolution and decentralisation of rights to biodiversity, the smaller are the chances for the monopolising tendencies to take hold.

Governments of the South can only be strengthened by standing behind their peoples and their biodiversity and supporting and protecting the democratic rights of diverse species to exist, and diverse communities to co-exist with them. If states in the South join the move to deny rights and to take away control over biodiversity from local communities, they too will be weakened and will lose their sovereign rights to and control over biodiversity to economic powers in the North whose global empires in the biotechnology era will be built on the destruction and colonisation of the South's biodiversity.

References

1. Shiva, V. 1989, *The Violence of the Green Revolution: Ecological Degradation and Political Conflict in Punjab*, p54

2. Mooney, P. R. 'The Law of the Seed: Another Development and Plant Genetic Resources', *Development Dialogue* 1983:1-2, p14

3. Treece, D. *Bound in Misery and Iron*, Survival International, p61

4. Ibid, p62

5. Jack Doyle, 1985, *Altered Harvest*, Viking, New York, p205

6. FAO, 1981, *Tropical Forest Resources*, FAO Forestry Paper 30, Rome

7. Raven, P. 1988, 'Our Diminishing Tropical Forests' in Wilson, E. O. (ed) *Biodiversity*, National Academy Press, Washington

8. Wilson, E. O. 1988 'The Current State of Biological Diversity' in Wilson, E. O. (ed) *Biodiversity*, National Academy Press, Washington

3 Biotechnology and the Environment

Introduction

THE FACT THAT ONE OF THE AGENDA items for the UN Conference on Environment and Development (UNCED) is the 'environmentally sound management of biotechnology' indicates that biotechnology is surrounded by social and ecological anxiety.

The first anxiety arises from the fact that the new biotechnologies tamper with the very fabric of life, and demand a fundamental restructuring of our minds, our ethics, our environmental, social and economic values and relationships. While biotechnology in its broadest sense is a very ancient group of technologies, it is the new biologies which generate new social, ecological, economic and political risks. The new biotechnologies consist of two major groups of technologies.

The first group, 'genetic engineering' refers to the new techniques deriving from advances in molecular biology, biochemistry and genetics. The second group is based on new cellular procedures based on the older technology of tissue culture.

Genetic engineering is a very powerful technique which theoretically allows any gene to be moved from any organism into another. Recombinant DNA technology has the potential to transform the genes into a global resource that can be used to shape novel life forms. It is this technical power which gives it the potential to become more pervasive than any technology in the past.

The new biotechnology has already found applications in primary industries (agriculture, forestry and mining), in secondary industries (chemicals, drugs, food) and in tertiary industries (health care, education, research, advisory services).

In addition to the wide ranging applications of biotechnology is the fact that the development of the new technologies is nearly entirely controlled by transnational enterprises, though universities and small firms evolved the techniques. These corporations are diversifying into every field of speciality which uses living organisms as a means of production. Traditional industry sectors are becoming less distinct and corporate boundaries virtually unlimited (Fowler et. al., 1988). This integration, centralisation and control carries with it an inherent destabilisation at the social, economic and ecological levels.

Biotechnology and Biohazards

(a) *The scientists' call for safety*

Technological innovation and scientific change do not merely bring benefits. They also carry social, ecological and economic costs.

It was the scientists closest to genetic engineering who first expressed concerns relating to the emergence of the new technology. In 1973, a group of prominent scientists called for a moratorium of certain types of research due to unknown risks and hazards associated with the possible escape and proliferation of novel forms of life. In 1975 at the Asiloniar Conference, part of the scientific community led by Paul Berg, a molecular biologist from Berkeley, attempted to agree on the need for regulation of biotechnological research (Krimsky, 1982).

Statement by scientists on potential biohazards of recombinant DNA molecules

Recent advances in techniques for the isolation and rejoining of segments of DNA now permit construction of biologically active recombinant DNA molecules in vitro. For example, DNA restriction endonucleases, which generate DNA fragments containing cohesive ends especially suitable for rejoining, have been used to create new types of biologically functional bacterial plasmids carrying antibiotic resistance markers and to link *Xenopus laevis* ribosomal DNA to DNA from a bacterial plasmid. This latter recombinant plasmid has been shown to replicate stably in *Escherichia coli* where it synthesizes RNA that is complementary to *X. laevis* ribosomal DNA. Similarly, segments of *Drosophila* chromosomal DNA have been incorporated into both plasmid and bacteriophage DNAs to yield hybrid molecules that can infect and replicate in *E. coli*.

Several groups of scientists are now planning to use this technology to create recombinant DNAs from a variety of other viral, animal, and bacterial sources. Although such experiments are likely to facilitate the solution of important theoretical and practical biological problems, they would also result in the creation of novel types of infectious DNA elements whose biological properties cannot be completely predicated in advance. There is serious concern that some of these artificial recombinant DNA molecules could prove biologically hazardous. One potential hazard in current experiments derives from the need to use a bacterium like *E. coli* to clone the recombinant DNA molecules and to amplify their number. Strains of *E. coli* commonly reside in the human intestinal tract, and they are capable of exchanging genetic information with other types of bacteria, some of which are pathogenic to man. Thus, new DNA elements introduced into *E. coli* might possibly become widely disseminated among human, bacterial, plant, or animal populations with unpredictable effects.

Concern for these emerging capabilities was raised by scientists attending the 1973 Gordon Research Conference on Nucleic Acids, who requested that the National Academy of Sciences give consideration to these matters. The undersigned members of a committee, acting on behalf of and with the endorsement of the Assembly of Life Sciences of the National Research Council on this matter, propose the following recommendations.

First, and most important, that until the potential hazards of such recombinant DNA molecules have been better evaluated or until adequate methods are developed for preventing their spread, scientists throughout the world join with the members of this committee in voluntarily deferring the following types of experiments.

● Type 1: Construction of new, autonomously replicating bacterial plasmids that might result in the introduction of genetic determinants for antibiotic resistance or bacterial toxin formation into bacterial strains that do not at present carry such determinants ; or construction of new bacterial plasmids containing combinations of resistance of clinically useful antibiotics unless plasmids containing such combinations of antibiotic resistance determinants already exist in nature.

● Type 2: Linkage of all or segments of the DNAs from oncogenic (cancer-inducing) or other animal viruses to autonomously replicating DNA elements such as bacterial plasmids or other viral DNAs. Such recombinant DNA molecules might be more easily disseminated to bacterial populations in huamns and other species, and thus possibly increase the incidence of cancer or other diseases.

Second, plans to link fragments of animal DNAs to bacterial plasmid DNA or bacteriophage DNA should be carefully weighed in light of the fact that many types of animal cell DNAs contain sequences common to RNA tumour viruses. Since joining of any foreign DNA to a DNA replication system creates new recombinant DNA molecules whose biological properties cannot be predicted with certainty, such experiments should not be undertaken lightly.

Third, the director of the National Institutes of Health is requested to give immediate consideration to establishing an advisory committee charged with (i) overseeing an experimental program to evaluate the potential biological and ecological hazards of the above types of recombinant DNA molecules; (ii) developing procedures which will minimise the spread of such molecules within human and other populations; and (iii) devising guidelines to be followed by investigators working with potentially hazardous recombinant DNA molecules.

Fourth, an international meeting of involved scientists from all over the world should be convened early in the coming year to review scientific progress in this area and to further discuss appropriate ways to deal with the potential biohazards of recombinant DNA molecules.

The above recommendations are made with the realisation (i) that our concern is based on judgements of potential rather than demonstrated risk since there are few available experimental data on the hazards of such DNA molecules and (ii) that adherence to our major recommendations will entail postponement or possibly abandonment of certain types of scientifically worthwhile experiments. Moreover, we are aware of many theoretical and practical difficulties involved in evaluating the human hazards of such recombinant DNA molecules. Nonetheless, our concern for the possible unfortunate consequences of indiscriminate application of these techniques motivates us to urge all scientists working in this area to join us in agreeing not to initiate experiments of types 1 and 2 above until attempts have been made to evaluate the hazards and some resolution of the outstanding questions has been achieved.

Paul Berg, Chairman
David Baltimore
Herbert W Boyer
Stanley N Cohen
Ronald W Davis
David S Hogness
Daniel Nathans
Richard Roblin
James D Watson
Sherman Weissman
Norton D Zinder

Committee on Recombinant DNA
Molecules Assembly of Life Sciences,
National Research Council,
National Academy of Sciences,
Washington, DC 20418

Later, as many scientists got involved in the commercial application of the new technologies – what Congressman Gore has called the 'selling of the tree of knowledge to Wall Street' – the self-criticism and self-restraint of the scientific community faded away.

The sustaining of the social impact analysis of the new technologies then became the responsibility of individual scientists and activists. The most persistent theme of the criticism has been the fear of adverse ecological and epidemiological consequences that might stem from the accidental or deliberate release of self-propagating genetically engineered organisms into the biosphere. Prominent scientists like Licbe Cavalieri, George Wald and David Suzuki have argued that the very power of the new technology outstrips our capacity to use it in safety, that neither nature's resilience nor our own social institutions are adequate protection against the unanticipated impacts of genetic engineering (Kloppenburg, 1988).

(b) *Public outcry against testing and deliberate release in the North*

(i) *The "ice minus" story*

Since frost damage is a major threat in the colder climate of the North, and runs up to $14 billion annually worldwide, biotechnologists are trying to make plants more tolerant to frost. They have isolated a gene which triggers ice nucleation in plant cells, and have deleted it from a certain bacterium called *Pseudomonous syringae*. The idea is that when this ice-minus bacteria is sprayed on a crop, such as Californian strawberries, it displaces the naturally occurring ice-forming bacteria, and the plants do not freeze when they normally would.

In 1983, Steven Lindow of Berkeley, and Advanced Genetic Sciences, a firm which was funding his work, were permitted by the National Institute of Health (NIH) Recombinant DNA Advisory Committee to run a field test. However, on September 4, a group of citizens and environmental interest groups based in Washington, DC – including Jeremy Rifkin, and the Foundation on Economic Trends, the Environmental Task Force, Environmental Action and the Humane Society – filed a suit against the NIH for approving the project. Among other things the suit charged that the NIH had not conducted an adequate assessment of the potential environmental risks of Lindow's field test and had 'been grossly negligent in its decision to authorise the deliberate release of the first genetically engineered life-forms'.

Among the risks that the public interest suit against NIH pointed to was the dramatic possibility that the frost-preventing bacteria might be swept into the upper atmosphere, disrupting the natural formation of ice-crystals, ultimately affecting local weather pattern and possibly altering the global climate. Eminent scientists like Eugene Odum and Peter Raven pointed to the ecological hazards of deliberate release of microorganisms since they reproduce rapidly, and their inter-relations with higher plants such as trees and plants are not known.

The public outcry associated with the ice-minus field test is pushing Northern governments and corporations to take their trials overseas to countries with little or no regulation, and that means the Third World.

(ii) *The BST story*

Bovine growth hormone, BST *(Bovine Somatotropin)* is the first hormone of the new biotechnological generation. Natural BST is a protein hormone that cows produce in sufficient quantities. In young animals it regulates muscle formation and growth, whereas in adult cows it controls milk production.

Genetically engineered BST is produced not by cows but by genetically engineered bacteria. Administered regularly to cows daily it increases milk yields by 7-14%.

Among the undesirable and negative side effects of biotechnological BST are severe deterioration of the health of the cow and higher surplus in regions where milk surpluses are already driving dairy farmers out of business. One estimate shows that if BST were licensed in the UK, by 1994-95 there would be 10% more dairy farmers going out of production than if it were not licensed. It is also not known whether hormone fragments will have side effects on the human body. There is no test for checking whether the growth hormone in cow's milk is natural or genetically engineered. There is no test for finding out what the recombinant version can do to the hormonal balance of people consuming BST-treated milk (Ram's Horn, 1991). In addition, the reduced immunity of the cow to disease will imply increased use of drugs and decreased quality of milk.

Animal rights activists, farmers and consumers in the North have achieved a ban on BST, such as in Wisconsin and Vermont in the USA. Three Canadian provinces have banned the selling of BST milk, and a national 'Pure Milk Campaign' has been launched to block the licensing of BST. The European Parliament supported a resolution calling for a worldwide ban on BST. BST has been banned in Denmark, Sweden and Norway.

In the USA a national coalition of farmers and consumers is organising a boycott of Mosanto, American Cynamid, Lily and Upjohn to prevent them from marketing BST.

(c) *Export of hazards to the Third World*

As bans and regulations delay tests and marketing in the North, biotechnology products will increasingly be tested in the

South to bypass regulation and public control.

The public, the scientists, and the official agencies of countries where these technologies are being developed, are aware of these hazards. Genetic engineering companies therefore face regulatory constraints, public protests and court injunctions domestically, and have started to conduct their release experiments involving recombinant organisms in countries where obstacles appear to be fewer due to laxer legislation and lower public awareness. As Dr Alan Goldhammer of the Industrial Biotechnology Association of the US had stated, 'the pathway may be clearer in foreign nations to getting approval'.

The Indian government has welcomed the biotech bandwagon of foreign companies by diluting the regulations and eroding the democratic structures that have existed within the country. The VAP (Vaccine Application Program) is clearly designed to by-pass safety regulations prevalent in the US because the memorandum of understanding states that all genetic engineering research 'will be carried out in accordance with the laws and regulations of the country in which the research is conducted'. Since India has no laws regulating genetic engineering, testing vaccines in India amounts to totally unregulated deliberate release.

The VAP was initiated in 1985 as part of the Reagan-Gandhi Science and Technology initiative, and the agreement was signed in Delhi on July 9, 1987. The project document states that 'The announcement of the VAP is an important recognition that vaccines are among the most cost-effective of health technologies, and their widespread use in both countries is key to controlling the burden of vaccine-preventable diseases.' The primary purpose of the project is to allow an extended range of trials of bio-engineered vaccines on animals and human subjects. The priority areas have been identified as cholera, typhoid fever, rotavirus, hepatitis, dysentery, rabies, pertussis, pneumonia and malaria, but these could change in succeeding years of the project as other areas of research opportunity are identified.

In 1986, the WISTAR institute based in Philadelphia hit the headlines for testing bio-engineered rabies vaccine on cattle in Argentina without the consent of the government or people of Argentina. When the Argentinian government became aware of the bovine rabies vaccine experiment in September 1986, it was immediately terminated. The Argentinian Ministry of Health alleged that farm hands who cared for the vaccinated cattle had been infected with the live vaccine.

WISTAR was driven out by the Argentinian government, but has been welcomed by the Indian government for participation in VAP. In fact, the project paper for VAP prepared by the US government applauds WISTAR for its achievements in the field of vaccine development, and specifically mentions the bovine rabies vaccine for field trials and other research.

The US government is evidently dictating the terms and conditions for these experiments, under VAP. The programme is financed by USAID (United States Agency for International Development) and the US Public Health Service. The total project cost is $9.6 million of which the US and Indian components are $7.6 million and $2 million respectively. Through the financial input the US government controls the agreement. Thus, all 'documents, plans, specifications, contracts, schedules and other arrangements with any modifications therein', must be approved by the USAID. On the other hand, scientists and scientific agencies in India directly concerned with the subject have been excluded from discussions on the programme.

(i) *Secrecy and the violation of Sovereignty*

The controversial Indo-US Vaccine project was signed bypassing the high-powered biotechnology scientific advisory committee set up by the Government of India. Dr Pushpa Bhargava, member of the committee and Director of the Centre for Cellular and Molecular Biology, said that the steps postulated in the Vaccine agreement 'are bound to come in the way of setting up our own research and development', and threaten to compromise Indian national sovereignty. The Union Science Minister, Mr K R Narayanan was not informed of the details of the agreement, nor was Dr V S Arunachalam, the Science Advisor to the Defence Minister. The Director General of the Indian Council of Medical Research stated categorically that he will not allow any vaccine to be tried on Indians unless the same is also approved for use in the US. As a result of scientific and public protest, the implementation of the programme has become even more secretive and totally removed from the public gaze.

A programme that will expose the Indian public to unknown hazards of live viruses used as vaccines denies the human subjects of the vaccine experiments the ethical right to prior informed consent. Human beings everywhere have a fundamental human right to know when they are being treated as guinea pigs, and they have a right to refuse to participate if they fear the exposure to unnecessary risks. With genetically engineered vaccines, the risks are indeed very high. Most researchers consider the use of attenuated lethal viruses as live vaccines too risky. Creating hybrid viruses has been viewed as one way to circumvent these risks. Recombinant DNA technology can be used to add the gene for an antigen of a lethal virus to the genome of a harmless virus, in an attempt to create a harmless living hybrid virus which, if used as a vaccine, provides immunity against the lethal virus. However, as Wheale and McNally report in *Genetic Engineering: Catastrophe or Utopia?*, recent research has shown that genetic manipulation of harmless viruses can turn them virulent. There is no 'safe' bio-engineered vaccine.

While VAP is totally irresponsible with regard to protection of people's health and environmental safety in the light of these hazardous implications, it shows great concern for the protection of corporate profits. It has a special clause for an accord on intellectual property which attempts to undo the public interest content of the Indian patent protection system.

Argentina and India are not the only countries to which bio-hazards are being exported. At a weeklong National Conference on Plant and Animal Biotechnology, in February 1990, USAID officials were pressing African states to allow field trials of geneti-cally altered organisms that might not be allowed in the regulatory systems in the North. Such was the concern that the Minister for Research, Science and Technology made a public pledge on the Conference's second day, stating that Kenya would not become a testing ground for dangerous new biotechnology products. Dr Calestus Juma, Director of the African Centre for Technology Studies (ACTS) advised scientists that the USAID is encouraging Third World countries in Asia and Latin America to undertake similar testing roles for private American firms. (African Diversity, June 1990).

(d) *Biohazards and Biosafety*

Ignorance about the ecological and health impacts of new technologies has far outweighed the knowledge needed for their production. As Jeremy Ravetz has stated, ignorance rather than knowledge characterises our times, and maintaining an ignorance about our ignorance is a central taboo of the technocratic culture (Ravetz, 1988).

It took 200 years of production based on fossil fuel before scientists realised that the burning of fossil fuels has unanticipated side effects – the destabilisation of the climate, the pollution of the atmosphere, and the creation of the greenhouse effect.

DDT was celebrated as an ultimate tool for ensuring public health. A Nobel Prize was awarded for its discovery. Today, DDT and other toxic pesticides are known to carry very high ecological and health costs, and many have been banned in the industrialised countries.

Union Carbide set up its chemical plants in India proudly announcing, 'We have a hand in India's future'. That future included the killing of 3,000 innocent people in December 1984 when MIC gas leaked from Carbide's pesticide plant in Bhopal.

Hazardous substances and processes have been manufactured faster than the structures of regulation and public control have evolved. We do not yet have full ecological criteria of testing for environmentally safe management of fossil fuel technologies of the mechanical enegineering revolution. The tests for environmentally safe management of the chemical engineering revolution are still in their infancy, leading to the marketing of products, processes and wastes which are proving to be ecologically unmanageable. Tests for safety in the genetic engineering revolution are yet to be conceived, since how the genetically modified life-forms interact with other organisms is totally unknown and uncharted territory.

Further, unlike hazardous chemicals such as pesticides and ecologically harmful substances like CFCs, the products of genetic engineering cannot be removed from the market. As George Wald has said in 'The case against Genetic Engineering': 'The results will be essentially new organisms, self perpetuating and hence permanent. Once created they cannot be recalled'.

(e) *Technology transfer and technology choice*

In biotechnology more than in any other area, lack of knowledge of hazards cannot be treated as safety. Restraint and caution is therefore considered the only wise strategy for unleashing powerful technologies with potentially serious risks in a context of near total ignorance.

For Third World countries, a special danger exists for being used as a testing ground and as guinea pigs. In addition, the uncertainties for the South are aggravated by the fact that the governments of the South want access to the new technologies of the North. In their haste to get access to the new biotechnologies, the Southern governments could unwittingly place themselves and their people and environment in this role of testing ground.

Therefore, to increase the benefits from the new technologies and to reduce their negative impacts, the Third World needs to rapidly evolve a framework of assessment of biotechnology on the basis of ecological, social and economic impact. Transfer of technology, an important issue for the South needs to be negotiated within such an assessment framework, so that socially desirable transfer of technology can take place while undesirable and hazardous transfer can be prevented.

In the area of the environmentally safe management of biotechnologies, it is important to have criteria of demarcation between technologies and products that are dangerous and unnecessary and those that are safe and desirable. This requires comparison and evaluation amongst different technology options, and the treatment of the new biotechnologies as merely one among many available alternatives to reach the same objective. In the final analysis, technology assessment and choice demands that technology be treated as what it is, a means, and not an end in itself.

Biotechnology and Chemical Hazards

While the area of biohazards is largely uncertain territory, after 40 years of chemical hazards, we know with certainty that human communities would rather live without them.

Will biotechnology lead to safer food, and food with fewer chemical residues from pesticides and other agrochemicals? Will the new biological approaches replace present agrochemicals?

As the Green Revolution miracle fades out as an ecological disaster, the biotechnology revolution is being heralded as an ecological miracle for agriculture. It is being offered as a chemical-free, hazard-free solution to the ecological problems created by chemically intensive farming. The past 40 years of chemicalisation of agriculture have led to severe environmental threats to plant, animal and human life. In the popular mind 'chemical' has come to be associated with 'ecologically hazardous'. The ecologically safe alternatives have been commonly labelled as 'biological'. Biotechnology has benefitted from its falling under the 'biological' category which has connotations of being ecologically safe. The biotechnology industry has described its agricultural innovations as 'ecology plus'.

It is, however, more fruitful to contrast the ecological with the engineering paradigm, and to locate biotechnology in the latter. The engineering paradigm offers technological fixes to complex problems, and by ignoring the complexity, generates new ecological problems which are later defined away as 'unanticipated side effects' and 'negative externalities'. Within the engineering ethos it is impossible to anticipate and predict the ecological breakdown that an engineering intervention can cause. Engineering solutions are blind to their own impacts. Biotechnology, as biological engineering, cannot provide the framework for assessment of its ecological impact on agriculture.

The first myth about biotechnology is that it is ecologically safe.

The second myth is that biotechnology will launch a period of chemical-free agriculture. However, most research and innovation in agricultural biotechnology are being undertaken by chemical multinationals such as Ciba Geigy, ICI, Monsanto and Hoechst.

The immediate strategy for these companies is to increase the use of pesticides and herbicides by developing pesticide and herbicide tolerant varieties. (Tables 1, 2 & 3). The dominant focus of research on genetic engineering is not on fertiliser-free and pest-free crops, but pesticide and herbicide resistant varieties. Twenty-seven corporations are working on virtually all major food crops to develop herbicide tolerance. For the seed-chemical multinationals this might make commercial sense, as Fowler et. al. point

Table 1						
Research Focus: Agricultural Biotechnology and the Private Sector: Company Activity by Region						
Type of Product	Number of Enterprises					
	USA	Canada	Europe	Latin America	Japan	Total
Seeds	137	14	38	3	11	203
Disease resistance	40	4	15	2	8	69
Herbicide resistance	26	3	8	0	1	38
Nitrogen-fixation	20	1	6	1	0	28
Pest resistance	18	2	4	0	0	24
Stress resistance	15	3	4	0	1	23
Protein improvement	18	1	1	0	1	21
Plant diagnostics	54	3	19	4	1	81
Plant food/feed	75	8	56	5	3	147
Other related products	10	2	12	25	1	50
Grand Total	276	27	125	37	16	481

Source: Manny Ratafia and Terry Purinton, Technology Management Group, World Agricultural Markets', Bio/Technology, March 1988, p.281.

Table 2

The Global Pesticides Industry: The Top Seven Enterprises in 1986 (US$ million)

Enterprise	State	Sales	Percentage of global sales	Herbicide tolerance
Bayer	FR Germany	2,344	13	Yes
Ciba-Geigy	Switzerland	2,070	12	Yes
ICI	UK	1,900	11	Yes
Rhone-Poulenc	France	1,500	9	Yes
Monsanto	USA	1,152	7	Yes
Hoechst	FR Germany	1,022	6	Yes
Du Pont	USA	1,000	6	Yes
Top Seven		10,988	64	

Source: *Development Dialogue: The Laws of Life*

Table 3

The Global Genetics Supply Industry: The Top Ten Enterprises in 1987 (US$ million)

Enterprise	State	Seed Sales	Percentage of global sales	Herbicide tolerance
Pioneer	USA	891.0	6.55	Yes
Shell	UK/ Netherlands	350.0	2.57	Yes
Sandox	Switzerland	289.8	2.13	Yes
Dekalb/Pfizer	USA	201.4	1.48	Yes
Upjohn	USA	200.0	1.47	Unknown
Limagrain	France	171.5	1.26	No
ICI	UK	160.0	1.19	Yes
Ciba-Geigy	Switzerland	152.0	1.12	Yes
Lafarge	France	150.0	1.10	Unknown
Volvo	Sweden	140.0	1.03	Unknown
Top Ten		2,705.7	19.89%	6 of 10

Source: *Development Dialogue: The Laws of Life*

out, since it is cheaper to adapt the plant to the chemical than to adapt the chemical to the plant. The cost of developing a new crop variety rarely reaches US$2 million whereas the cost of a new herbicide exceeds US$40 million. Herbicide and pesticide resistance will also increase the integration of seeds/chemicals and the control of TNCs in agriculture.

A number of major agrichemical companies are developing plants with resistance to their brand of herbicides. Soyabeans have been made resistant to Ciba-Geigy's Atrazine herbicides, and this has increased annual sales of the herbicide by US$120 million. Research is also being done to develop crop plants resistant to other herbicides such as Dupont's 'Gist' and 'Glean' and Monsanto's 'Round-Up', which are lethal to most herbaceous plants and thus cannot be applied directly to crops. The successful development and sale of crop plants resistant to brand name herbicides will result in further economic concentration of the agro-industry market, increasing the market power of transnational companies.

For the Third World farmer this strategy for employing more toxic chemicals on pesticide and herbicide resistant varieties is suicidal, in a lethal sense. Thousands of people die annually as a result of pesticide poisoning. In 1987, more than 60 farmers in India's prime cotton growing area of Prakasam district in Andhra Pradesh committed suicide by consuming pesticide because of debts incurred for pesticide purchase. The introduction of hybrid cotton created pest problems. Pesticide resistance resulted in epidemics of white-fly boll worm, for which the peasants used more toxic and expensive pesticides, incurring heavy debts and thus being driven to suicide (Ramprasad, 1988). Even when pesticides and herbicides do not kill people, they kill people's sources of livelihoods. The most extreme example of this destruction is that of bathua, an important green leafy vegetable with very high nutritive value which grows as an associate of wheat. However, with intensive chemical fertiliser use bathua has become a major competitor of wheat and has been declared a 'weed' that is killed with herbicides and weedicides.

Herbicide resistance also excludes the possibility of rotational and mixed-cropping, which are essential for a sustainable and ecologically balanced form of agriculture, since the other crops would be destroyed by the herbicide. US estimates now show a loss of US$4 billion per annum due to crop loss as a result of herbicide spraying. The destruction in the Third World will be far greater because of higher plant diversity, and the prevalence of diverse occupations based on plants and biomass. Thousands of rural women who make their living by weaving baskets and mats with wild reeds and grasses are losing their livelihoods because increased use of herbicides is killing the reeds and grasses. The introduction of herbicide-resistant crops will increase herbicide use and thus increase the damage to economically and ecologically useful plant species.

Strategies of genetic engineering for herbicide resistance which are destroying useful species of plants can also end up creating superweeds. There is an intimate relationship between weeds and crops, especially in the tropics where weedy and cultivated varieties have genetically interacted over centuries and hybridise freely to produce new varieties. Genes for herbicide tolerance, pest-resistance, stress-tolerance that genetic engineers are striving to introduce into crop plants may be transferred to neighbouring weeds as a result of naturally occurring gene transfer. (Wheale and McNally, 1988, p.172).

The Third World needs to ban the introduction of herbicide and pesticide resistant crops because of their health impact, ecological impact and economic impact, including labour displacement and the increase of capital intensity of agriculture.

Biotechnology and Biodiversity

There is a prevalent misconception that biotechnology development will automatically lead to biodiversity conservation. The main problem with viewing biotechnology as a miracle solution to the biodiversity crisis is related to the fact that biotechnologies are, in essence, technologies for the breeding of uniformity in plants and animals. Biotechnology corporations do talk of contributing to genetic diversity. As John Duesing of Ciba-Geigy states, 'Patent protection will serve to stimulate the development of competing and diverse genetic solutions with access to these diverse solutions ensured by free market forces at work in biotechnology and seed industries'. However, the 'diversity' of corporate strategies and the diversity of life-forms on this planet are not the same thing, and corporate competition can hardly be treated as a substitute for nature's evolution in the creation of genetic diversity.

Corporate strategies and products can lead to diversification of **commodities**; they cannot enrich nature's diversity. This confusion between commodity diversification and biodiversity conservation finds its parallel in raw-material diversification. Although breeders draw genetic materials from many places as raw material input, the seed commodity that is sold back to farmers is characterised by uniformity. Uniformity and monopolistic seed supplies go hand in hand. When this monopolising control is achieved through the molecular mind, destruction of diversity becomes more accelerated. As Jack Kloppenburg has warned, 'Though the capacity to move genetic material **between** species is a means for introducing additional variation, it is also a means for engineering genetic uniformity **across** species.'

The application of DNA transfer to crop improvement may result in a greater degree of genetic uniformity among cultivars. Calgene has isolated a bacterial gene which can be transferred to a tobacco plant and when successfully expressed confers resistance to the herbicide glyphosate (Monsanto's 'Round-Up').

It might be said that Calgene has added variability to the tobacco genepool. But if that gene is a commercial success and is incorporated into most tobacco cultivars, the result may be increased genetic uniformity in that crop (Kloppenburg, 1988). And it was the broad distribution of a single genetic character that led to the corn blight epidemic of 1970 in the US.

Tissue culture will also generate uniformity in agriculture and forestry. Companies such as Shell, Weyerhaeuser and International Paper are looking to mass produce genetically identical seedlings. Normally, populations of organisms are diverse. Some have a capacity to resist a disease, say a fungus, while others do not. Diversity allows a species to survive. However, should cloned genetically uniform trees prove susceptible to a pathogen or pest, millions of acres of forest and years of production might be lost.

Biotechnology may well diminish genetic diversity and increase genetic vulnerability (Yoxen 1986, Kloppenburg 1988).

Most large scale plantations of commercial species are now being introduced in the Third World. Shell has been given 60,000 ha in Uruguay, with financing by the World Bank. Large areas of land in Thailand have also been taken over for tree planting by Shell. If future plantations financed through the Tropical Forestry Action Plan use clones of eucalyptus and tropical pines, the ecological and financial costs of collapse will be borne by Third World countries, in addition to existing costs of biodiversity destruction, and displacement of local communities. (Lohmann, 1991).

Shifting the focus on Biodiversity and
Biotechnology Linkages

1. Two thirds of the world's biodiversity lies in the Third World.
2. The most significant relationship between biodiversity and biotechnology
    is that the former acts as a raw material base for biotechnology based
    industries located in the North and in the private sector.
3. The focus on biodiversity discussions has so far been confined to Box 1.
    The focus should, in fact, be on Box 2 which allows the evolution of
    coherent and integrated framework to discuss issues such as control over
    biological resources, conservation of biodiversity, technology transfer,
    financial mechanisms, codes of conduct on impact of the new technolo-
    gies on health and environmental safety, economic impacts on
    employments and commodity exports.

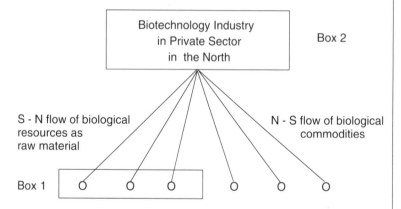

Biotechnology Industry
in Private Sector
in the North

Box 2

S - N flow of biological
resources as
raw material

N - S flow of biological
commodities

Box 1

Biodiversity in the Third World

Focus on Box 1 in global negotiations threatens to erode the rights of local
custodians of biodiversity and sovereignty of national governments in the
Third World and shifts the control of the South's biodiversity to the North while
neglecting the political, economic, ecological challenges raised by the new
biotechnologies. Focus on Box 2 allows an integrated and comprehensive
discussion of all aspects of biodiversity and biotechnology while preserving
the sovereignty of local communities and national governments.

Biotechnology Substitutes and Economic Displacement in the Third World

Probably the most serious impact of biotechnology will be the displacement of some agricultural export commodities from the Third World, with its related impacts on the national economy and employment. Plant tissue culture offers increased possibilities of substituting specialities with industrially produced inputs. Many high value plant-derived products used for pharmaceuticals, dyes, flavourings and fragrances are vulnerable to displacement as a result of current research (OCED, 1989).

The impact of successful production of substitutes will be felt most by countries which have, in an earlier international division of labour, been made dependent upon exports of the natural products concerned. This will particularly be destructive to economies in Africa which depend entirely on single crops for most of their export earnings. While historically Africa was used to grow crops needed for Europe, in the emerging world order based on new biotechnologies, Africa will become dispensable as the North finds biotech substitutes for African crops.

When factories close in the North, compensation is given to workers. When crops first introduced by global agribusiness are displaced by technologies developed by agribusiness, the small peasant and agricultural worker are left to fend for themselves, as are their countries. The South needs to develop an agenda for compensation which is based on a notion of historical justice and which can be tabled before the full deployment of the new biotechnologies which are being developed to reduce dependence on the Third World (Hobbelink, 1991, Fowler et. al., 1988).

Table 4: An Illustrative List of Plant Tissue Culture Research Activities

Plant cultured	Plant products to be cultured	Plant country of origin	Research Organisa-tion	Value of produce per kg (US$)	Market size (US$) million
Lithos-permum	Shikonin	Korea,China	Mitsui Petro-chemical (Japan)	4500	
Pyrethrum	Pyrethrines	Tanzania Ecuador,India	University of Minne-sota	300	20 (US)
Papaver	Cadeine opium	Turkey		850	50 (US)
Sapota	Chicle	Central America	Lotte (Japan)		
Catha-rantus	Vincristine		Canadian National Research Council	5000	18-20 (US)
Jasminum	Jasmine	Many producers		5000	0.5 (World)
Digitalis	Digitoxine-digoxine		Univ. of Tubingen Bochringer-Mannheim (Germany)	3000	20-55 (US)
Cinchona	Quinine	Indonesia	Plant Science Ltd. (UK)		
Cocoa	Cocoa butter	Brazil,Ghana	Cornell University Hershey, Nestle		891 (World)
Thauma-tococcus	Thaumatine	Liberia,Ghana Malaysia	Tate and Lyle (UK)		
Rauwolfia	Reserpine				80 (US)

Source: *Kenney and Buttel,F., 'Biotechnology: Prospects and Dilemmas for Third World Development', Development and Change, Vol.16, No.1, 1985.*

Biotechnology, Privatisation and Concentration

Most of the adverse impacts of biotechnology are related to the fact that the new technologies are evolving under control of the transnational private sector.

Biotechnology was born in the laboratories of universities and other public research institutions. Some scientists then moved out to build their own biotechnology companies. Now it is the giant agrochemical, pharmaceutical and food processing transnationals that dominate the research and the markets.

Along with the privatisation trend is the trend towards con-centration. As Henk Hobbelink has put it, 'the few are becoming fewer, and the big grow bigger' (Hobbelink, 1991). Where 30 manufacturers were involved in pesticide development in the mid-1970s in the US, there are only a dozen today.

For decades, the top 30 drug producers have remained the same. Today, 10 corporations control 28% of the world market as a result of mergers.

TNCs are buying up most of the seed companies. Today, the top ten companies control over 20% of the global market, and have interests in chemicals, pesticides and pharmaceuticals. It is ex-pected that by 2000 the top ten will control most of the seed market, including that controlled by farmers who save their own seed, and that controlled by the public agriculture research system which played a central role in the development and distribution of Green Revolution seed varieties.

The privatisation trends are clear in changes in India's seed policy. They are also being indicated in China, which has been a pioneer in developing hybrid rice, which is capable of boosting harvests by 25%. But the rice variety that allows its production –

a so-called male-sterile line of rice that will not self-seed – is not being distributed in Asia. Two multinational companies – Cargill Seeds and Occidental Petroleum's Ring Around Products – are known to have exclusive license agreements with the Chinese government for seed development, production and marketing in specified countries.

An agreement between the Chinese government and the two US companies forbade the sharing of information and materials concerning hybrid rice with other governments or with IRRI. The Chinese government was therefore forced to withdraw its support for an IRRI hybrid rice training course (Robert Walgate, 1990).

The erosion of a public system of control and regulation is therefore inevitable with growing privatisation. The OCED survey of industry and government of main issues raised by biotechnology shows that markets are the main driving force for industry.

The divergence between the imperative for private profits and people's well being is expected to grow. Corporations will attempt to adjust society to their need for profits. They will increasingly use the state to restructure the relationships between people and between North and South to suit their needs. The issue of privatisation is increasingly becoming a threat to democracy and people's will, as the same scientists work on contract for TNCs, function on government regulatory bodies, and dominate scientific research. In this context it is up to citizens, free of TNC and government control, to keep public issues and priorities alive, and have a space for public control of the new biotechnologies.

Biotechnology, Patents, and Private Property in Life Forms

The ultimate expression of privatisation of biotechnology is the desperate urge by TNCs, operating through the US Trade

Representative, World Bank, GATT and WIPO, to have a uniform patent system that allows them to own all life on this planet as their private property.

Patents in the context of agriculture and food production involve ownership over life forms and life processes. Monopoly ownership of life creates an unprecedented crisis for agricultural and food security, by transforming biological resources from commons into commodities. It also generates a crisis of values and ends which guide social organisation, technological change, and development priorities.

Development debates throughout the South have revealed that development is not a neutral category. Development for some implies underdevelopment for many. This is as true for agricultural development as development in other spheres, as experience with the Green Revolution has shown. Patents in agriculture are being pushed by Northern governments and corporations as essential for agricultural development in the Third World. Dupont's Ralph Hardy has said: 'The competitive position of US industry in biotechnology would be improved if there were international conventions that would provide greater uniformity with respect to patentability and property rights. There are some countries that do not recognise (property rights) and this will significantly retard the development and early commercialisation of products that would improve the health and food supply of these countries.'

Yet it is not the countries who are supposed to benefit from worldwide intellectual property protection who are demanding patent protection. It is the multinational corporations.

Monsanto's Nicholas Riding echoes Dupont's Hardy when he says: 'The major challenge to genetic engineering scientists and companies as well as national governments is to support uniform worldwide property rights.'

This is just another way of stating that global monopoly over agriculture and food systems should be handed over as a right to multinational corporations. With worldwide patent protection, agribusiness and the seed trade are trying to achieve truly global reach. While the rhetoric is agricultural development in the Third World, the enforcement of strong patent protection for monopoly ownership of life processes will undermine and underdevelop agriculture in the Third World in a number of ways.

Firstly, it will undermine our cultural and ethical fabric based on agriculture, in which the fundamental life processes are treated as sacred, not as commodities to be bought and sold on the market. The sacred cow will give way to patented livestock, and, according to the US patent law, the offspring of patented livestock would also be subject to royalty charges throughout the 17 or 22 years of patent protection. Fowler et. al. have called this the patent version of the old 'original sin'. It will affect not just animals but plants as well. Seeds, which have been treated as sacred, as gifts exchanged freely between farmers, will become patented commodities. Hans Leenders, Secretary General of the world seed houses and their breeders, has proposed to abolish the farmers' right to save seed. He says: 'Even though it has been a tradition in most countries that a farmer can save seed from his own crop, it is under the changing circumstances not equitable that a farmer can use this seed and grow a commercial crop out of it without payment of a royalty.....the seed industry will have to fight hard for a better kind of protection.'

The corporate demand to change a common heritage into a commodity and to treat profits generated through this transformation as a property right will lead to erosion not just at the ethical and cultural level, but also at the economic level for Third World farmers. The Third World farmer has a three level relationship with the corporations demanding a monopoly of life forms and life processes. Firstly, the farmer is a **supplier** of germ plasm to TNCs. Secondly, the farmer is a **competitor** in terms of innovation and rights to genetic resources. Finally, the Third World farmer is a **consumer** of the technological and industrial products of TNCs. Patent protection displaces the farmer as a competitor, transforms him into a supplier of free raw material, and makes him totally dependent on industrial supplies for vital inputs like seeds. Above all, the frantic cry for patent protection in agriculture is for protection **from** farmers, who are the original breeders and developers of biological resources in agriculture. It is argued that patent protection is essential for innovation – however, it is essential only for innovation that brings profits to corporate business. Farmers have carried out innovations over centuries and public institutions have carried out innovations over decades without any property rights or patent protection.

Further, unlike plant breeders rights (PBR), the new utility patents are very broad based, allowing monopoly rights over individual genes and even characteristics. PBR is not an ownership over germ plasm in the seeds; it gives only a monopoly right for the selling and marketing of a specific variety. The monopoly rights of industrial patents go much further. They allow for multiple claims that can cover not only whole plants but plant parts and processes as well. So, according to attorney Anthony Diepenbrock: 'You could file for protection of a few varieties of crops, their macroparts (flowers, fruits, seeds and so on) their micro parts (cells, genes, plasmids, and the like) and whatever novel processes you develop to work with these parts all using one multiple claim.'

Patent protection implies the exclusion of farmers' rights over resources having those genes and characteristics. This will undermine the very foundation of agriculture in India. For example, a recent patent has been granted to Sungene for a sunflower variety with very high oleic acid content. The claim allowed was for the characteristic itself (i.e. high oleic acid), and not just the genes producing the characteristic. Sungene has notified others involved in sunflower breeding that the development of any variety high in oleic acid will be considered an infringement.

In the 1985 judgement in Ex parte Hibberd, molecular genetics scientist Kenneth Hibberd and his co-inventors were granted patents on tissue culture seed and whole plant of a corn line selected from tissue culture. The Hibberd application included over 260 separate claims, which give the molecular genetics scientists the right to exclude others from use of all 260 aspects. While apparently Hibberd provides a new legal context for corporate competition, the most profound impact will be felt in the competition between farmers and the seed industry. As Kloppenburg has indicated, with Hibberd, a judicial framework is now in place that may allow the seed industry to realise one of its longest held and most cherished goals: to force all farmers into dependence on the companies every year. Industrial patents allow the right to **use** the product, not to **make** it. Since seed makes itself, a strong utility patent for seed implies that a farmer purchasing seed would have the right to use (to grow) the seed, but not the right to make seed (to save and replant). The farmer who saves and replants seeds of a patented plant variety will be in violation of the law.

These processes of outlawing the original custodians of plant genetic resources will happen slowly. But patent protection is central to transnational agricultural interest, which makes quite clear that it is their monopoly on markets rather than the development of farmers of the South that is at issue.

Patents and intellectual property rights are the remaining hurdles to be crossed for large scale distribution of biotech seeds by transnational corporations. For instance, one of the clauses of the new seed policy in India directs all companies importing seeds to make a small quantity available to the gene bank of the government controlled National Bureau of Plant Genetic Resources (NBPGR). The corporate giants are, of course, unwilling to accept that clause and want its removal. As Jan Nefkins, General Manager of Cargill South-east Asia Limited, points out: 'No company would be willing to part with what they took years and spent millions of dollars developing. It's a question of intellectual property rights.'

Intellectual property rights in the agricultural sector will marginalise not just the farmers but also the national research and breeding system which has been built-up so carefully. Public systems are the second competitor which private industry needs to displace since they work on plant breeding on the basis of public interest, not the private profit motive. All kinds of pressures are already underway to reduce the role of public institutions in agricultural research and development. Process patents achieve this even without a policy decision. Patents on breeding techniques have long been anathema to public breeders, because their purpose is to exclude others from access to the means to perform research. What is called into question is not ownership of the product, but the right to 'do science'. With broad based patent claims, large areas of research can become corporate monopoly, as in the case of US patent No. 4,326,358 issued in 1982 to Agrigenetics Research Associates. Patent 4,326,358 made 14 separate claims, but in its essentials it gave Agrigenetics Research Associates rights to the process of using clonally propogated parental lines to develop new hybrid plant varieties.

While patents are supposed to stimulate innovation, they actually stifle it. One outcome of property rights over living systems has been secrecy in plant breeding and plant genetics research, and restrictions on the exchange of germ plasm. Secrecy for patents and exclusivity together will 'choke off' all scientific exchange in plant genetics. In addition, rather than simply stimulating innovation, the patent system applied to living matter redirects attention towards those products that provide for the broadest and easiest patent protection, not the largest public good.

As Jack Doyle observes: 'Patents and intellectual property rights are more a statement of territory than a measure of innovation. But when those "territories" are food and the substances behind the production of food, there is an order of protected power that is broad and far reaching.'

India's patent laws have excluded the monopoly over biological processes which are essential to sustenance and survival. Food, plant and animal varieties, and biological processes for the production of plants and animals are excluded from patentability.

However, living organisms are central to production processes in biotechnology. The need for ownership rights to living organisms is essential for the next stage of capital accumulation by global corporations. At the core of the business are property rights and patents which guarantee profits, excluding others from rights and access to the means of survival.

Royalty payments from farmers are expected to reap $7 billion for patent holders. Arguing that these potential profits are not being made because of an absence of a worldwide application of US patent law, the US is claiming that it is losing $100 to $300 billion. The Third World is already collapsing under debt and interest payments leading to a $50 billion transfer of funds from the poor South to the rich North. Additional revenues generated for Northern corporations by proprietary rights on living resources is an impossible demand, though a typically outrageous one.

The accusation of 'piracy' that the US is making against Third World countries is actually more applicable to the US itself. On a recent estimate based on a seven-country study, the US claimed that its corporations were losing US$135 million a year in royalty payments for agricultural chemicals, and US$1.7 billion in pharmaceuticals. Extrapolated to the entire Third World, this royalty loss is US$202 million for agricultural chemicals and US$2.5 billion for pharmaceuticals.

The Rural Advancement Fund International has shown that if Third World contributions are taken into account, the 'pirate' roles are substantially reversed. The US owes the Third World US$302 million for royalty for farmers' seeds and US$5.1 billion for pharmaceuticals. In other words, in these two biological industry sectors, the US owes US$2.7 billion to the South.

Who owes whom is a tricky issue, especially when it comes to profiteering from biological resources which have originated in the Third World and which continue to provide sustenance and survival to millions of farmers. These resources have also reproduced freely and have been commonly accessible to all to support livelihood and nutrition needs. Indian patent law has excluded private ownership of the biological foundations of agriculture to ensure that entitlements to food and nutrition are as broad-based as possible. Transnational corporations, with the help of GATT, the World Bank, and instruments like Super 301 of the US Trade Act, are demanding the inclusion of these living resources into frameworks of patents and private ownership. Such an inclusion will exclude our rights to survival as a country and as a people. Sovereignty in the matter of patent law is essential because it is a matter of survival, especially for the economically weaker sections of our society which have no purchasing power and can be protected only through public interest, not by the profit motive. The choice is clear. It is protection of life vs. the protection of profits.

Appendices

Table 1: Top ten pesticides houses

(1988 sales, US$ billion adapted for recent take-overs)

	Pesticide Sales	% of Global Market
1. Ciba-Geigy (CH)	2.14	10.70
2. Bayer (FRG)	2.07	10.37
3. ICI (UK)	1.96	9.80
4. Rhone-Poulenc (F)	1.63	8.17
5. Du Pont (USA)	1.44	7.19
6. Dow Elanco (USA)*	1.42	7.11
7. Monsanto (USA)	1.38	6.89
8. Hoechst (FRG)	1.02	5.12
9. BASF (FRG)	1.00	5.00
10. Shell (NL/UK)	0.94	4.69
Total top ten	15.00	75.02
Global Sales	20.00	100.00

*The pesticides divisions of Dow Chemicals and Eli Lilly (Elanco) merged into Dow Elanco in 1989. Other major pesticide companies include (with 1988 pest. sales): Schering (US$0.75 billion), American Cyanamid (USA, US$0.69 billion), Sandoz (CH US$0.60 billion) and Kumiai (J, US$0.46 billion).

Source: *AGROW, 'Ciba-Geigy still number one in 1988', No. 92, 28 July 1989, p1.*

Cited in Henk Hobbelink, *Biotechnology and the Future of World Agriculture*, Zed Press, 1991.

Table 2

The top ten pharmaceuticals corporations
(1987 sales in US$ billion, adapted for recent take-overs)

	Pharma. Sales	% of Global Market
Merck (US)	4.23	3.53
SmithKline-Beecham (US/UK)	4.00	3.33
Bristol M. - Squibb (US)	3.90	3.25
Hoechst (FRG)	3.51	2.93
Glaxo (UK)	3.37	2.81
R. Poulenc-Rorer (F/US)	3.30	2.75
Ciba-Geigy (CH)	3.17	2.64
Bayer (FRG)	2.96	2.47
Am. Home Products (US)	2.93	2.44
Sandoz (CH)	2.75	2.29
Total top ten	34.12	28.43
Global Market	120.00	100.00

Table 3

Top ten seed corporations
(1988 sales, US$ million, adapted for recent take-overs)

	Seed Sales	% of Global Market
1. Pioneer Hi-Bred (US)	735	4.90
2. Sandoz (CH)*	507	3.38
3. Limagrain (F)	370	2.46
4. Upjohn (USA)	280	1.87
5. Aritrois (F)*	257	1.71
6. ICI (UK)	250	1.67
7. Cargill (USA)	230	1.53
8. Shell (NL/UK)	200	1.33
9. Dekalb-Pozer (USA)	174	1.16
10. Ciba-Geigy (CH)	150	1.00
Total top ten	3,098	20.65
Global	15,000	100.00

* Sandoz bought Hilleshog from Volvo (S), thus substantially increasing seed sales (up from an estimated US$290 million in 1987). Aritois is a new joint venture in which Rhone Poulenc and Lafarge-Coppee are bringing their seeds interests together. The new group includes Clause, the French market leader in vegetable and ornamental seeds. One study puts seeds sales of Ciba-Geigy as high as US$245 million and Aritois sales as low as US$104 million. Shell sold off part of its seed interests to Limagrain, thus adding some US$100 million to the French company's revenues. Other recent take-overs: Limagrain bought Shissler Seed Co. (USA), ICI bought Contiseed from Continental Grain (USA), Cargill bought Canola Corp., Unilever bought PBI (UK) and Barenburg (NL). Other major seed companies include: KWS (FRG), Lubrizol (USA), Takii (J), Cebecco (NL), Elf Aquitaine (Sanofi - F).

Sources: *'Les chimistes tentent de se constituer de nouveaux bastions sur le marché mondial des semences', in Le Monde, Paris, 21 November 1989 (based on ICI estimates); Rhone-Poulenc/Lafarge-Coppee seed joint venture', AGROW, Richmond UK, No. 95, 8 September 1989; several other issues of AGROW were used.*

References

1. Cary Fowler, et.al (1988) *Laws of Life*, Development Dialogue, Dag hammarsjold Foundation, Uppsala

2. Sheldon Krimsky, (1982) *Genetic Alchemy: The Social History of the Recombinant DNA Debate*, MIT Press,Cambridge, MA

3. Peter Wheale and Ruth McNally, (1988) *Genetic Engineering: Catastrophe or Utopia,* Harvester, 1988

4. OECD, 1989 *Biotechnology – International Trends and Perspectives*

5. Henk Hobbelink, (1991) *Biotechnology and the Future of World Agriculture,* ZED Books, London

6. Robert Walgate, (1990) *Biotechnology*, Earthscan, London

7. Hans W Leenders, (1986) *Reflections on 25 years of service to the International Seed Trade Federation*, Seeds Men's Digest 37:5 (May): 8-9

8. Jack Doyle, (1985) *Altered Harvest: Agriculture,Genetics and the Fate of the World's Food Supply*, Viking, NewYork

4 The Seed and the Spinning Wheel: Technology Development and Biodiversity Conservation

Introduction

BIODIVERSITY CONSERVATION is generally conceived of as independent of the technologies of production which use and transform biological resources. This Chapter demonstrates the interdependence of these two factors. In Third World countries where most of the world's biodiversity is concentrated, many communities of tribals and peasants derive their livelihoods and satisfy multiple needs directly from the rich diversity of biological resources. Production technologies based on uniform monocultures of trees, crops or livestock threaten livelihoods while they displace biodiversity. The common misconception that diversity is linked to low productivity, and that uniformity is essential for high productivity is addressed. It is shown that when multiple yields, values and outputs of biological systems are taken fully into account, diversity does not preclude high productivity. Using the symbol of Gandhi's spinning wheel, the paper urges for a deeper consideration of the social and ecological context in which technology development takes place.

Biological richness is not uniformly distributed across the world. It is concentrated in the tropical countries of the Third World. Most current conservation plans for biodiversity, however, originate in the North, and they carry within them the social categories of development and planning that are characteristic of industrialised and affluent countries.

According to the dominant paradigm of production, diversity goes against productivity, which creates an imperative for uniformity and monocultures. This has generated the paradoxical situation in which modern plant and animal improvement has been based on the destruction of the biodiversity which it uses as raw material. The irony of plant and animal breeding is that it destroys the very building blocks on which the technology depends. Forestry development schemes introduce monocultures of industrial species like eucalyptus and push into extinction the diversity of local species which fulfil local needs. Agricultural modernisation schemes introduce new and uniform crops into farmers' fields and destroy the diversity of local varieties. In the words of Professor Garrison Wilkes of the University of Massachusetts, this is analogous to taking stones from a building's foundations to repair the roof (Myers 1985).

In agriculture and forestry, in fisheries and animal husbandry, production is being incessantly pushed in the direction of destruction of diversity. Production based on uniformity thus becomes the primary threat to biodiversity conservation and to sustainability.

However, this threat to biodiversity from technology development has been little understood and analysed. This Chapter is an attempt to fill this lacunae; and to enrich the understanding of the relation between technology, natural resources and human needs. In particular it focuses on the social and ecological impact of technology change in areas related to biological resources.

Technology Development and Sustainability

Science and technology are conventionally accepted as what scientists and technologists produce, and development is accepted as what science and technology produce. Scientists and technologists are in turn taken to be that sociological category formally trained in Western science and technology, either in institutions or organisations in the West,or in Third World institutions mimicking the paradigms of the West. These tautological definitions are unproblematic if one leaves out people, especially poor people; if one ignores ecological and cultural diversity and distinct civilisational and natural histories of our planet which have created diverse and distinctive cultures and ecosystems. Development in this view is taken as synonymous with the introduction of Western science and technology in non-Western contexts. The magical identity is development = modernisation = Westernisation.

In a wider context, where science is viewed as 'ways of knowing' and technology as 'ways of doing', all societies, in all their diversity, have had science and technology systems on which their distinct and diverse development has been based. Technologies or systems of technologies bridge the gap between nature's resources and human needs. Systems of knowledge and culture provide the framework for the perception and utilisation of natural resources. Two changes occur in this shift of definition of science and technology. Science and technology are no longer viewed as uniquely Western but as a plurality associated with all cultures and civilisations. And a particular science and technology do not automatically translate into development everywhere. Ecologically and economically inappropriate science and technology can become causes of underdevelopment and poverty, not solutions to underdevelopment and impoverishment. Ecological inappropriateness is a mismatch between the ecological processes of nature which renew life support systems and the resource demands and impacts of technological processes. Technological processes can lead to higher withdrawals and consumption of natural resources or higher additions of pollutants than ecological limits allow. In such cases they contribute to underdevelopment through destruction of ecosystems.

Economic inappropriateness is the mismatch between the needs of society and the requirements of a technological system. Technological processes create demands for raw materials and markets, and control over both raw materials and markets becomes an essential part of the politics of technological change.

The lack of the theoretical cognition of the two ends of technological processes, their beginning in natural resources and their end in basic human needs, has created the current paradigm for economic and technological development which demands increasing withdrawals of natural resources and generates increasing addition of pollutants while marginalising and dispossessing an increasing number of people from the productive process. These characteristics of contemporary scientific industrial development are the primary causes for the ecological, political and economic crisis. The combination of ecologically disruptive scientific and technological modes and the absence of the criteria for evaluating scientific and technological systems, in terms of efficient resource use and capability of satisfying basic needs, has created conditions where society is increasingly propelled towards ecological and economic instability and has no rational and organised response to arrest and curtail these destructive tendencies (Shiva et al. 1991).

In Third World economies, many communities depend on biological resources for their sustenance and well being. In society, biodiversity is simultaneously a means of production, and an object of consumption. It is the survival base that has to be conserved. Sustainability of livelihoods is ultimately connected to the conservation and sustainable use of biological resources in all their diversity.

However, biodiversity-based technologies of tribal and peasant societies have been viewed as backward and primitive and have been displaced by technologies which use biological resources in such a way that they destroy diversity and people's livelihoods.

Diversity and Productivity

There is a general misconception that diversity-based production systems are low productivity systems. However, the high productivity of uniform and homogeneous systems is a contextual and theoretically constructed category; based on taking only one-dimensional yield and outputs into account. On this same construct, the multi-dimensional yields of diversity-based systems of production are excluded and ignored. Were the 'miracle' seeds of the Green Revolution inherently superior and 'advanced' in comparison to the diversity of indigenous crops and varieties that they displaced? The miracle of the new seeds has most often been communicated through the term 'high yielding varieties' (HYVs). As analysed in Chapter I, the HYV category is a central category of the Green Revolution paradigm. Unlike what the term suggests, there is no neutral or objective measure of 'yield' on the basis of which the cropping systems based on miracle seeds can be established to be higher yielding than the cropping systems they replace.

The HYV category is similarly not a neutral observational concept. This was explained earlier in Chapter I. Its meaning and measure is determined by the theory and paradigm of the Green Revolution, which is not easily and directly translatable for comparison with the agricultural concepts of indigenous farming systems for a number of reasons. The Green Revolution category of the HYV is essentially a reductionist category which decontextualises properties of both the native and the new varieties. Through the process of decontextualisation, costs and impacts are externalised and systematic comparison with alternatives is precluded.

Cropping systems, in general, involve an interaction between soil, water and plant genetic resources. As an example, in indigenous agriculture, cropping systems include a symbiotic relationship between soil, water, farm animals and plants. Green Revolution agriculture replaces this integration at the level of the farm with the integration of inputs such as seeds and chemicals. The seed/ chemical package sets up its own interactions with soils and water systems, which are, however, not taken into account on the assessment of yields.

Chapter I gives details of how modern plant-breeding concepts like HYV reduce farming systems to individual crops and parts of crops. Crop components of one system are then measured with crop components of another. Since the Green Revolution strategy is aimed at increasing the output of a single component from a farm, at the cost of decreasing other components and increasing external inputs, such a partial comparison is by definition biased to make the new varieties 'high yielding' even when at the systems level, they may not be. Traditional farming systems are based on mixed and rotational cropping systems of cereals, pulses, oilseeds with diverse varieties of each crop, while the Green Revolution package is based on genetically uniform monocultures. No realistic assessment has ever been made of the yield of the diverse crop outputs in the mixed and rotational systems. Usually the yield of a single crop like wheat or maize is singled out and compared to yields of new varieties.

The measurement of output is also biased by restricting it to the marketable part of crops. However, in a country like India, crops have traditionally been bred and cultivated to produce not just food for man but fodder for animals, and organic fertiliser for soils.

Table 5 in Chapter I illustrates how diverse the grain and straw ratio of different rice varieties can be.

In the breeding strategy for the Green Revolution, multiple uses of plant biomass seem to have been consciously sacrificed for a single use, with non-sustainable consumption of fertiliser and water. The increase in marketable output of grain has been achieved at the cost of decrease of biomass for animals and soils and the decrease of ecosystem productivity due to the over-use of resources. The increase in production of grain for marketing was achieved in the Green Revolution strategy by reducing the biomass for internal use on the farm. Similarly, indigenous breeds of Indian cattle have been described as 'unproductive', and all animal-breeding programmes are aimed at replacing them with exotic strains of the Jersey, Holstein, Friesian, Red Dane and Brown Swiss. However, Indian cattle do not only yield milk. They provide energy and fertiliser, which are crucial to farming systems.

It should be noted that two-thirds and more of the power requirements of Indian villages are met by some 80 million work animals of which 70 million are the male progeny of what the Western perspective sees as 'useless' low milk-yielding cows. It has been calculated that to replace animal power in agriculture, India would have to spend about US$1,000 million annually on petrol. Indian cattle excrete 700 million tonnes a year of recoverable manure: half of this is used as fuel, liberating the thermal equivalent of 27 million tonnes of kerosene, 35 million tonnes of coal or 68 million tonnes of wood, all of which are scarce resources in India; the remaining half is used as fertiliser. As for other livestock produce, it may be sufficient to mention that the export of hides, skins, etc. brings in $150 million annually into the national coffers. With limited resources, indigenous cattle produce a multiplicity of uses (George 1985).

Indian cattle provide food in excess of the edible food consumed, in contrast to the US where six times as much edible food is fed to cattle as is obtained from them (Leon 1975) (Table 1). Yet this highly efficient food system, based on the multiple uses of

cattle, has been dismantled in the name of efficiency and 'development' by the reductionist strategies of the green and white revolutions splitting and dichotomising an integrated system of crop
production and animal husbandry, necessary for maintaining
each other sustainably (Shiva 1988, 1991).

The low productivity of diverse, multi-dimensional systems
and the high productivity of uniform, one-dimensional systems of
agriculture, forestry and livestock is therefore not a neutral, scientific measure but is biased towards the commercial interests for
whom maximising of one-dimensional output is an economic
imperative.

Table 1						
Inputs and useful outputs from US cattle and Indian cattle and buffalo (1972). Reprinted from Leon (1975).						
Inputs and outputs	Matter (10^{10}kg)		Energy (10^{12}calories)		Protein (10^{9}kg)	
	USA	India	USA	India	USA	India
Inputs						
Edible by man	11.9	0.68	38.6	1.7	16.0	2.1
Inedible by man	22.2	40.00	88.0	120.5	25.1	33.3
Total	34.1	40.68	126.6	122.2	41.1	35.4
Outputs						
Work	-	-	-	6.50	-	-
Milk	1.12	0.51	5.04	2.09	2.06	0.88
Meat	0.90	0.50	4.40	2.23	0.17	0.11
Hides	0.11	0.07	-	-	-	-
Manure	0.87	10.81	-	16.16	-	-
Total	3.00	11.89	9.44	26.98	2.23	0.99
Efficiency (%)	9	29	7	22	5˙	3

This push towards uniformity, however, undermines the diversity of biological systems which form the production system. It also undermines the livelihoods of the people whose work is associated with diverse and multiple use systems of forestry, agriculture and animal husbandry. As an example, in the State of Kerala in India, which derives its name from the coconut palm, coconut is cultivated in a multi-storeyed, high-intensity cropping system along with betel and pepper vines, bananas, tapioca, drumstick, papaya, jackfruit, mango and vegetables. Compared to annual labour requirement of 157 man-days ha^{-1}, in a monoculture of coconut palm, the mixed cropping system increases employment to 960 man-days ha^{-1} (Government of Kerala 1964).

In the dry land farming systems of the Deccan, the shift from mixed-cropping of millets with pulses and oilseeds to eucalyptus monocultures has led to a loss of employment of 250 man-days ha^{-1} yr^{-1} (Shiva and Bandyopadhyay 1987).

When labour is scarce and costly, labour displacing technologies are productive and efficient. When labour is abundant, labour displacement is unproductive because it leads to poverty, dispossession and destruction of livelihoods.

In Third World situations, sustainability has therefore to be achieved at two levels simultaneously – the sustainability of natural resources and the sustainability of livelihoods. Biodiversity conservation has therefore to be linked to conservation of livelihoods derived from biodiversity.

Conservation of the Seed and the Spinning Wheel

The conservation of livelihoods along with the conservation of resources has been a special concern for us in India. It was the basis of our freedom movement and the struggle against colonialism.

Mahatma Gandhi had recognised that poverty and underdevelopment in India was rooted in the destruction of jobs linked to our rich textile industry. The regeneration of livelihoods was central to the process of regaining independence. Gandhi categorically stated that what is good for one nation situated in one condition is not necessarily good for another differently situated. One man's food is often another man's poison. Mechanisation is good when hands are too few for the work intended to be accomplished. But according to Gandhi, it is an evil where there are more hands than required for the work as is the case in India (Pyarelal 1959).

The spinning wheel became for Gandhi and India a symbol of a technology that conserves resources, people's livelihoods and people's control over their livelihoods. In contrast to the imperialism of the British textile industry, which had destroyed India's industrial base, the 'charkha' was decentred and labour generating, not labour displacing. It needed people's hands and minds, instead of treating them as surplus, or as mere inputs into an industrial process.

This critical mixture of decentralisation, livelihood generation, resource conservation and strengthening of self-reliance was essential to undo the waste of centralisation, livelihood destruction, resource depletion and creation of economic and political dependence that had been engendered by the industrialisation associated with colonialism.

Gandhi's spinning wheel is a challenge to notions of progress and obsolescence that arise from absolutism and false universalism in concepts of science and technology development. Obsolescence and waste are social constructs that have both a political and ecological component. Politically, the notion of obsolescence gets rid of people's control over their lives and livelihoods by defining productive work as unproductive and removing people's control over production in the name of progress. It would rather waste hands than waste time. Obsolescence also destroys the regenerative capacity of nature by substituting manufactured uniformity in place of nature's diversity. Technological obsolescence translates into obsolescence of biodiversity. This induced dispensability of poorer people on the one hand and diversity on the other constitutes the political ecology of technological development guided by narrow and reductionist notions of productivity. Parochial notions of productivity, perceived as universal, rob people of control over their means of reproducing life and rob nature of her capacity to regenerate diversity.

Ecological erosion and destruction of livelihoods are linked to one another. Displacement of diversity and displacement of people's sources of sustenance both arise from a view of development and growth based on uniformity created through centralised control. In this process of control, reductionist science and technology act as handmaidens for economically powerful interests. The struggle between the factory and the spinning wheel continues as new technologies emerge for the manipulation of biological resources.

As the spinning wheel was rendered backward and obsolete by the industrialisation of the manufacture of textiles, farmers' seeds are being rendered obsolete and valueless by technological change associated with the industrialisation of seed production.

The indigenous varieties or landraces in agriculture have evolved through millenia of natural and human selection. These varieties produced and used by Third World farmers worldwide are called 'primitive cultivars'. Those varieties created by modern plant breeders in international research centres or by transnational seed corporations are called 'advanced' or 'elite'. The hierarchy in words like 'primitive' and 'elite' has strong cultural roots, even when these words are used in scientific domains. Underlying these categorisations is an inherent bias which assumes that

technologies that emerge in the industrialised North are superior
in an absolute sense. However, the experience of the Green
Revolution informs us that in the domain of biodiversity, technology development can lead to progress for one interest group
while creating underdevelopment for others.

The change in the nature of seed is justified by creating a
framework that treats self-regenerative seed as 'primitive' and as
'raw' germplasm and the seed that is inert without inputs and non-
reproducible as a finished product. The whole is rendered partial,
the partial is rendered whole. The commoditised seed is ecologically incomplete and ruptured at two levels:

1. It does not **reproduce** itself, while by definition, seed is
    a regenerative resource. Genetic resources are thus,
    through technology, transformed from a renewable into
    a non-renewable resource.
2. It does not **produce** by itself. It needs the help of inputs
    to produce. As the seed and chemical companies merge,
    the dependence on inputs will increase, not decrease.
    And ecologically, whether a chemical is added exter-
    nally or internally, it remains an external input in the
    ecological cycle of the reproduction of seed.

It is the shift from the ecological processes of reproduction to
the technological processes of production that underlies both the
problem of dispossession of farmers and of genetic erosion.

The new plant biotechnologies will follow the path of the
earlier HYVs of the Green Revolution in pushing farmers onto a
technological treadmill. Biotechnology can be expected to increase
the reliance of farmers on purchased inputs even as it accelerates
the process of polarisation. It will even increase the use of chemicals
instead of decreasing it. The dominant focus of research in genetic
engineering is not on fertiliser-free and pest-free crops, but pesticide

and herbicide-resistant varieties. For the seed-chemical multinational companies, this might make commercial sense, since it is cheaper to adopt the plant to the chemical than to adopt the chemical to the plant. The cost of developing a new crop variety rarely reaches US$2 million, whereas the cost of a new herbicide exceeds US$40 million (Fowler et al. 1988).

Like Green Revolution technologies, biotechnology in agriculture can become an instrument for dispossessing the farmer of seed as a means of production. The relocation of seed production from the farm to the corporate laboratory relocates power and value between the North and South; and between corporations and farmers. It is estimated that the elimination of home grown seed would dramatically increase the farmers' dependence on biotech industries by about US$6,000 million annually (Kloppenburg 1988).

It can also become an instrument of dispossession by selectively removing those plants or parts of plants that do not serve the commercial interest but are essential for survival of nature and people. 'Improvement' of a selected characteristic in a plant, is also a selection **against** other characteristics which are useful to nature, or for local consumption. Improvement is not a class or gender neutral concept. Improvement of partitioning efficiency is based on the enhancement of the yield of the desired product at the expense of unwanted plant parts. The desired product is, however, not the same for rich people and poor people, or rich countries and poor countries, nor is efficiency. On the input side, richer people and richer countries are short of labour and poorer people and poorer countries are short of capital and land. Most agricultural development, however, increases capital input while displacing labour, thus destroying livelihoods. On the output side, which parts of a farming system or a plant will be treated as 'unwanted' depends on what class and gender one is. What is unwanted for the better off may be the wanted part for the poor. The plants or 'plant parts' which serve the poor are the ones whose supply is squeezed by the normal priorities of improvement in response to commercial forces.

Conclusion

The destruction of people's livelihood and sustenance goes hand in hand with the erosion of biological resources and their capacity to fulfil diverse human needs while regenerating and renewing themselves. Attempts to increase commodity flows in one direction generate multiple levels of scarcities in related outputs. Increase of grain leads to decrease of fodder and fertiliser. Increase of cereals leads to decrease of pulses and oilseeds. The increase is measured. The decrease goes unnoticed, except by those who are deprived through the creation of new scarcity. Both people and nature are impoverished; their needs are no longer met by the one-dimensional production systems which replace biologically rich and diverse ecosystems, and put added burdens on remaining pockets of biodiversity to satisfy their needs.

The extinction of people's livelihoods and sustenance is closely connected with the erosion of biodiversity. Protection of biodiversity can only be ensured by regenerating diversity as a basis of production in agriculture, forestry and animal husbandry. The practice of diversity is the key to its conservation.

Biodiversity cannot be conserved until diversity is made the logic of production. If production continues to be based on the logic of uniformity and homogenisation, uniformity will continue to displace diversity. 'Improvement' from the corporate viewpoint or from the viewpoint of Western agricultural research is often a loss for the Third World, especially the poor in the Third World. There is therefore no inevitability that production acts against diversity. Uniformity as a pattern of production becomes inevitable only in a context of control and profitability.

All systems of sustainable agriculture, whether of the past or the future, work on the basis of the perennial principles of diversity and reciprocity. The two principles are not independent but interrelated. Diversity gives rise to the ecological space for give and take, for mutuality and reciprocity. Destruction of diversity is linked to the creation of monocultures, and with the creation of monocultures, the self-regulation and decentred organisation of diverse systems gives way to external inputs and external and centralised control. Sustainability and diversity are ecologically linked because diversity offers the multiplicity of interactions which can heal ecological disturbance to any part of the system. Non-sustainability and uniformity means that disturbance to one part is translated into a disturbance to all other parts. Instead of being contained, ecological destabilisation tends to be amplified. Closely linked to the issue of diversity and uniformity is the issue of productivity. Higher yields and higher production have been the main push for the introduction of uniformity and the logic of the assembly line. The imperative of growth generates the imperative for monocultures. Yet this growth is, in large measure, a socially constructed, value-laden category. It exists as a 'fact' by excluding and erasing the facts of diversity and production through diversity. Sustainability, diversity and decentred self-organisation are therefore linked, as are unsustainability, uniformity and centralisation.

Diversity as a pattern of production, not merely of conservation, ensures pluralism and decentralisation. It prevents the dichotomising of biological systems into 'primitive' and 'advanced', or of knowledge systems into 'primitive' and 'advanced'. As Gandhi challenged the false concepts of obsolescence and productivity in the production of textiles by his search for the spinning wheel, groups across the Third World are challenging the false concepts of obsolescence in agricultural production which necessarily generate non-sustainability. They are searching for the diversity of seeds used by farmers over centuries and making them the basis of a futuristic, self-reliant, resilient and sustainable agriculture (Altieri 1991; Shiva 1991).

References

1.   Altieri, M. 1991. Traditional farming in Latin America. Ecologist 21;
      93-96.

2.   Fowler, C., E. Lachkovics, P. Mooney and H. Shand. 1988. The laws
      of life: Another development and the New Biotechnologies.
      Development Dialogue 1988 (1-2): 1-350.

3.   George, S. 1985. *Operation Flood*. Oxford University Press, New
      Delhi.

4.   Government of Kerala. 1984. Report of High Level Committee on
      Land and Water Resources. Trivandrum, India.

5.   Kloppenburg, J. 1988. *First the Seed*. Cambridge University Press,
      Cambridge.

6.   Kuhn, T. 1972. *The Structure of Scientific Revolution*. University of
      Chicago Press, Chicago.

7.   Leon, B. 1975. Agriculture: A sacred cow. Environment 17:38.

8.   McNeely, J. A., K. R. Miller, W. V. Reid, R. A. Mittermeier and T. B.
      Werner. 1990. *Conserving the World's Biological Diversity*. IUCN,
      Gland, Switzerland.

9.   Myers, N. (ed) 1985. *The Gaia Atlas of Planet Management*. PAN
      Books, London.

10.   Pearse, A. 1980. *Seeds of Plenty, Seeds of Want*. UNRISD, Geneva.

11.   Pyarelal. 1959. *Towards New Horizons*. Navjivan Press, Ahmedabad.

12.   Shiva, V. 1988. *Staying Alive: Women, Ecology and Development*. Kali
      for Women, New Delhi and Zed Books, London.

13.   Shiva, V. 1991. *Violence of the Green Revolution*. Third World Net-
      work, Penang and Zed Books, London.

14.   Shiva, V. and J. Bandyopadhyay. 1987. *Ecological Audit of Eucalyptus
      Cultivation*. Research Foundation for Science, Technology and
      Natural Resource Policy, Dehra Dun.

15.   Shiva, V., J. Bandyopadhyay, P. Hedge, B. V. Krishnamurthy, J.
       Kurien, G. Narendranath, V. Ramprasad and S. T. S. Reddy. 1991.
       *Ecology and the Politics of Survival: Conflicts Over Natural Resources in
       India*. UNU, Tokyo and SAGE, New Delhi/Newbury Park, London.

16.   Yegna Iyengar, A. K. 1944. *Field Crops of India*. BAPPCO, Bangalore
       (reprinted 1980).

5 The Biodiversity Convention: an Evaluation from the Third World Perspective

THE BIODIVERSITY CONVENTION started out primarily as an initiative of the North to 'globalise' the control, management and ownership of biological diversity (which due to ecological reasons lies primarily in the Third World) so as to ensure free access to the biological resources which are needed as 'raw material' for the biotechnology industry.

It was, however, in the interest of the North to keep access to biodiversity delinked from access to biotechnology, and to focus only on the international regulation of biodiversity conservation. The issue of the regulation of biotechnology was not present in any of the drafts of the convention until July 1991 .

However, after the Preparatory Committee meetings of the United Nations Conference on Environment and Development in Geneva in August 1991, the links between biodiversity negotiations and biotechnology issues were strengthened and sections were introduced on biosafety and the need for regulation of biotechnology. This was, in part, a result of intensive interaction between the Group of 77 countries and the Third World Network, which held regular briefing sessions for delegates.

The Convention was starting to shape up into a reflexive document, in which both biodiversity and biotechnology, and both the North and the South would be internationally regulated. It was with these diverse elements that the convention draft went

for the final international negotiation committee meeting to Nairobi.

President Bush's announcement that he will not sign the
Biodiversity Convention at the Earth Summit has probably been
the most significant event related to the UN Conference on Environment and Development. Governments and NGOs, media ce-
lebrities have all been putting pressure on Bush to go along with
other countries in signing the biodiversity treaty.

The US's refusal is based on grounds that the text is 'seriously
flawed'. From the ecological perspective it is true that the text is
flawed but this is not the flaws Bush is indicating. The flaws have
in fact been introduced by the US in the final negotiations in
Nairobi and relate in particular to issues of patenting and intellectual
property rights. According to Bush, the convention is not strong
enough on patents. This is however merely a ploy to wrest further
concessions from the South. In fact the convention is too strong on
patents, and too weak on the intellectual and ecological rights of
indigenous peoples and local communities. The text produced in
Nairobi is not fully satisfactory from the citizens' point of view.
Among the weaknesses in the current text is a basic flaw in Article
3 which states that:

*States have in accordance with the Charter of the United Nations and
the Principles of international law, the sovereign right to exploit their
own resources pursuant to their environmental policies, and the responsibility to ensure that activities within their jurisdiction or control
do not cause damage to the environment of other states or of areas beyond
the limits of national jurisdiction.*

What is missing in the principle is the sovereign right of local
communities which have conserved and maintained biodiversity,
and whose cultural survival is linked intimately to the survival of
biodiversity, to conserve and use biological diversity. It is ironical

that a convention for the protection of biodiversity has been distorted into a convention to exploit it.

Another flaw in the convention is the assumption that biotechnology is essential for the conservation and sustainable use of biological diversity: as stated in Article 16(1). Diverse species exist independent of technology, though, biotechnology depends on biodiversity to provide raw material for commercial objectives. Unlike other commodities, biotechnology commodities replace and substitute the original biodiversity which they consume as raw material.

It is this double transformation induced by biotechnology that has significant impact on the Third World. Not only is biodiversity devalued from being a 'means of production' into being mere raw material, it is also displaced by the genetically uniform biotechnology products. It is essential to remember that new biotechnologies are essentially technologies for the production of uniformity.

The third flaw in the biodiversity convention is that it has accepted patents in the area of living resources. The clauses on patents have been introduced only in the final round of negotiations in Nairobi. Article 17 para 2 and 3 of the 20 February draft addressed the issue of transfer of technology on fair and concessional terms, with no commitment to patents and intellectual property protection. The final draft of the convention had introduced a clause that stated that 'In the case of technology subject to patents and other intellectual property rights, such access and transfer shall be provided on terms which recognise and are consistent with the adequate and effective protection of intellectual property rights.'

While the US has been successful in introducing patents into the Biodiversity Convention, it is still unhappy with the language, in particular that IPRs are impediments rather than a prerequisite for technology transfer. The US is also unhappy with Article 16(5) which states:

Contracting parties, recognising that patents and other IPRs may influence on the implementation of this convention, shall cooperate in this regard, subject to national legislation and international law in order to ensure that such rights are supportive of and do not run counter to its objectives.

Other last minute changes that the US manipulated in the Biodiversity Convention in Nairobi is the exclusion of the world's crop gene banks. By not including the issue of ownership and related rights over genetic resources presently in gene banks, the Biodiversity Convention could result in serious economic loss to developing countries as industrial countries (where most of the gene banks are located) can be expected to rush to patent these genetic materials.

Experts in international public organisations who have been closely following the recent developments in biodiversity negotiations warn that if the Convention comes into force, industrial country governments would take legislative measures to enable the patenting of genetic materials presently located in gene banks in their countries.

Much of these materials had been collected from developing countries by international agricultural research institutes, and two-thirds of all seeds collected in gene banks are in industrial countries or are stored in international research centres controlled by Northern countries and the World Bank.

The ownership of these genetic materials has not been internationally clearly defined, since much of them was collected with international public finance and their origin is mainly in developing countries whilst the location of the gene banks is in the North.

This lack of clarity over ownership and rights discouraged Northern governments and the international research centres from patenting the genetic materials. However this may now change if the Biodiversity Convention comes into force.

The reason is that the Convention deals only with access to genetic resources to be collected in future, whilst excluding the hundreds of thousands of samples now housed in gene banks or botanical gardens. There is thus to be no internationally binding obligation on these gene banks or botanical gardens to pay the countries of origin of the genetic resources, or to share equitably with them the benefits of the use of the materials and the technology.

At a meeting on 22 May 1992, the CGIAR (Consultative Group on International Agricultural Research) announced its intention to patent some of their genetic materials and to collaborate with private corporations in the use of the materials. Northern governments can also be expected to introduce legislation (if necessary) to enable the patenting of genetic materials in gene banks, even though they had been collected from developing countries with the understanding that they would be easily or freely available for public use. The gene banks are now planning to screen the many thousands of samples to select the useful genes for patenting. The patented materials could then be made available to biotechnology companies and to others (including Third World farmers), who have to pay royalties. These royalties would not flow to the real countries of origin of the genetic resources (the South); indeed the South would ironically have to pay to get access to them.

After the patenting, developing countries would not have the legal right to the genetic materials even if the genetic resources had been collected from these countries with international public financing by CGIAR centres or by the IBPGR (the International

Board for Plant Genetic Resources which is associated with a United Nations agency but controlled by CGIAR).

The gene banks, most of them in or under the control of Northern countries, hold some 90% of known genetic resources of the most important agricultural crops. The exclusion of these valuable materials from obligatory rules is thus a fundamental flaw of the Biodiversity Convention. Worse, the Convention, by being silent on how these materials should be treated, now opens the door to Northern countries to patent the resources in their gene banks. They could argue that since the major international regulatory instrument (the Convention) does not cover the issue of rights and obligations over genetic resources in gene banks and botanical gardens, then governments are free to introduce their own laws and regulations enabling intellectual rights protection over these materials.

The result of the consequent patenting is that developing countries would have to pay high prices for seeds and genetic materials in these gene banks and for modified genetic materials. At the same time they would not be compensated for the knowledge of their farmers and forest peoples, which is the source of the evolutionary use of the seeds and other materials in agricultural production. The Biodiversity Convention does not recognise the right of informal innovators (including farmers) to be compensated.

A fifth flaw in the Biodiversity Convention is in last minute changes in the definitions. Terms such as 'country of origin', 'in situ conditions' and 'eco-system' have been so defined as to lend themselves to convenient interpretations in order to suit the interests of the North. International experts are of the view that the Biodiversity Convention, if adopted and brought into force, could eventually prove to have facilitated the opening of a floodgate of patenting genetic resources presently in gene banks, as well as endowing rights to controllers of gene banks and botanical gardens (situated in the North), similar to the rights of countries of origin where the resources naturally developed, that is the South. If that happens, then the convention would eventually prove disadvantageous to the economic interests of developing countries.

A sixth weakness of the biodiversity convention is that it has accepted the World Bank's Global Environmental Facility as the interim financial mechanism. An independent funding mechanism, called the Biological Diversity Fund in earlier drafts on which the Third World was insisting, has been sacrificed.

On patents, access to genetic resources, access to technology and financial mechanism, we have systematically lost ground in biodiversity.

In the light of the above weaknesses, which have been highlighted by the Third World Network experts at Rio, appearances notwithstanding, the Biodiversity Convention has the risk of favouring the US more than it favours the Third World. Much will depend on future interpretations and amendments. Probably the only aspect that the US would further like to dilute are the clauses on biosafety in Article 19. This article was introduced after the Third Prepcom in Geneva. It was diluted during the final negotiations in Nairobi in May 1992. In Articles 19(3) and 19(4) which were Articles 20(3) and 20(4) in the fifth draft of 20 February 1992, reference to the more accurate term 'genetically modified organisms'(GMOs) has been removed and substituted by the vague term 'living modified organism resulting from biotechnology'. In spite of this dilution in terminology, the biosafety clauses have survived. The issue of biosafety and regulation of biotechnology is a major reason for the Bush decision to not sign the treaty.

The recent record of the US has been a systematic dismantling of the regulatory framework for ensuring environmental and health security in the area of biotechnology. The regulations of the Food and Drug Administration (FDA) have been drastically weakened. Instead, the White House Council on Competitiveness headed by Dan Quayle has urged all agencies to speed up clearances for genetically engineered products. As a result, the US Department of Agriculture approved 22 field tests of genetically modified organisms in the period from 20 March to 21 April 1992. Federal regulations which include those concerning biotechnology control were suspended in a 90-day moratorium, which was further extended by another four months on 29 April 1992.

More recently, the FDA has ruled that food products altered by genetic engineering raised no new or unique safety issues and will be regulated no differently than food created by conventional means. Thus foods which have had genes from animals introduced into them are to be treated 'as natural' and 'safe' on the basis that the transferred gene occurs naturally in the original organism. Already, human genes have been transferred to pigs, and chicken genes to crops. In such instances, complex ecological, ethical, cultural and religious problems can emerge which have been totally ignored and even suppressed.

Quite clearly, Article 14 of the Biodiversity Convention which addresses biosafety issues would make it necessary to examine safety in biotechnology and render the ongoing deregulation in the US illegal under international law. On the other hand the convention would strengthen regulation related to people's health and environmental safety. This clause which protects the environment and human lives is what Mr Bush calls a 'serious flaw' since he is openly committed to the cause of industry.

The Bush Administration did not want the Earth Summit to put in place any international safety regulation on biotechnology industry, but insisted on patent regulation to protect industry profits.

It wants to give industry a guarantee that they will have the licence to experiment and manipulate life forms, under patent protection without any ethical, social or environmental responsibility.

Commentators have called the Biodiversity Convention 'legalised theft'. What is at stake for us is the very foundation of our life support and our civilisation. Third World governments need to ensure that amendments and interpretations of the convention are made so that the survival of our diverse communities and the diverse species with which they live is not sacrificed. For us in the Third World, the protection of plants is predicated on the protection of people who have been their custodians throughout history. It is this partnership between living biodiversity and living communities which the Biodiversity Convention needs to conserve.

Appendix

Convention on Biological Diversity
5 June 1992

Preamble

The Contracting Parties,

Conscious of the intrinsic value of biological diversity and of the ecological, genetic, social, economic, scientific, educational, cultural, recreational and aesthetic values of biological diversity and its components,

Conscious also of the importance of biological diversity for evolution and for maintaining life sustaining systems of the biosphere,

Affirming that the conservation of biological diversity is a common concern of humankind,

Reaffirming that States have sovereign rights over their own biological resources,

Reaffirming also that states are responsible for conserving their biological diversity and for using their biological resources in a sustainable manner,

Concerned that biological diversity is being significantly reduced by certain human activities,

Aware of the general lack of information and knowledge regarding biological diversity and of the urgent need to develop scientific, technical and institutional capacities to provide the basic understanding upon which to plan and implement appropriate measures,

Noting that it is vital to anticipate, prevent and attack the causes of significant reduction or loss of biological diversity at source,

Noting also that where there is a threat of significant reduction or loss of biological diversity, lack of full scientific certainty should not be used as a reason for postponing measures to avoid or minimize such a threat,

Noting further that the fundamental requirement for the conservation of biological diversity is the *in-situ* conservation of ecosystems and natural habitats and the maintenance and recovery of viable populations of species in their natural surroundings,

Noting further that *ex-situ* measures, preferably in the country of origin, also have an important role to play,

Recognizing the close and traditional dependence of many indigenous and local communities embodying traditional lifestyles on biological resources, and the desirability of sharing equitably benefits arising from the use of traditional knowledge, innovations and practices relevant to the conservation of biological diversity and the sustainable use of its components,

Recognizing also the vital role that women play in the conservation and sustainable use of biological diversity and affirming the need for the full participation of women at all levels of policy-making and implementation for biological diversity conservation,

Stressing the importance of, and the need to promote, international, regional and global cooperation among States and intergovernmental organizations and the non-governmental sector for the conservation of biological diversity and the sustainable use of its components,

Acknowledging that the provision of new and additional financial resources and appropriate access to relevant technologies can be expected to make a substantial difference in the world's ability to address the loss of biological diversity,

Acknowledging further that special provision is required to meet the needs of developing countries, including the provision of new and additional financial resources and appropriate access to relevant technologies,

Noting in this regard the special conditions of the least developed countries and small island States,

Acknowledging that substantial investments are required to conserve biological diversity and that there is the expectation of a broad range of environmental, economic and social benefits from those investments,

Recognizing that economic and social development and poverty eradication are the first and overriding priorities of developing countries,

Aware that conservation and sustainable use of biological diversity is of critical importance for meeting the food, health and other needs of the growing world population, for which purpose access to and sharing of both genetic resources and technologies are essential,

Noting that, ultimately, the conservation and sustainable use of biological diversity will strengthen friendly relations among States and contribute to peace for humankind,

Desiring to enhance and complement existing international arrangements for the conservation of biological diversity and sustainable use of its components, and

Determined to conserve and sustainably use biological diversity for the benefit of present and future generations,

Have agreed as follows:

Article 1. Objectives

The objectives of this Convention, to be pursued in accordance with its relevant provisions, are the conservation of biological diversity, the sustainable use of its components and the fair and equitable sharing of the benefits arising out of the utilization of genetic resources, including by appropriate access to genetic resources and by appropriate transfer of relevant technologies, taking into account all rights over those resources and to technologies, and by appropriate funding.

Article 2. Use of Terms

For the purposes of this Convention:

'*Biological diversity*' means the variability among living organisms from all sources including, *inter alia*, terrestrial, marine and other aquatic ecosystems and the ecological complexes of which they are part; this includes diversity within species, between species and of ecosystems.

'*Biological resources*' includes genetic resources, organisms or parts thereof, populations, or any other biotic component of ecosystems with actual or potential use or value for humanity.

'*Biotechnology*' means any technological application that uses biological systems, living organisms, or derivatives thereof, to make or modify products or processes for specific use.

'*Country of origin of genetic resources*' means the country which possesses those genetic resources in *in-situ* conditions.

'*Country providing genetic resources*' means the country supplying genetic resources collected from *in-situ* sources, including populations of both wild and domesticated species, or taken from *ex-situ* sources, which may or may not have originated in that country.

'*Domesticated or cultivated species*' means species in which the evolutionary process has been influenced by humans to meet their needs.

'*Ecosystem*' means a dynamic complex of plant, animal and micro-organism communities and their non-living environment interacting as a functional unit.

'*Ex-situ conservation*' means the conservation of components of biological diversity outside their natural habitats.

'*Genetic material*' means any material of plant, animal, microbial or other origin containing functional units of heredity.

'*Genetic resources*' means genetic material of actual or potential value.

'*Habitat*' means the place or type of site where an organism or population naturally occurs.

'*In-situ conditions*' means conditions where genetic resources exist within ecosystems and natural habitats, and, in the case of domesticated or cultivated species, in the surroundings where they have developed their distinctive properties.

'*In-situ conservation*' means the conservation of ecosystems and natural habitats and the maintenance and recovery of viable populations of species in their natural surroundings and, in the case of domesticated or cultivated species, in the surroundings where they have developed their distinctive properties.

'*Protected area*' means a geographically defined area which is designated or regulated and managed to achieve specific conservation objectives.

'*Regional economic integration organization*' means an organization constituted by sovereign States of a given region, to which its member States have transferred competence in respect of matters governed by this Convention and which has been duly authorized, in accordance with its internal procedures, to sign, ratify, accept, approve or accede to it.

'*Sustainable use*' means the use of components of biological diversity in a way and at a rate that does not lead to the long-term decline of biological diversity, thereby maintaining its potential to meet the needs and aspirations of present and future generations.

'*Technology*' includes biotechnology.

Article 3. Principle

States have, in accordance with the Charter of the United Nations and the principles of international law, the sovereign right to exploit their own resources pursuant to their own environmental policies, and the responsibility to ensure that activities within their jurisdiction or control do not cause damage to the environment of other States or of areas beyond the limits of national jurisdiction.

Article 4. Jurisdictional Scope

Subject to the rights of other States, and except as otherwise expressly provided in this Convention, the provisions of this Convention apply, in relation to each Contracting Party:

(a) In the case of components of biological diversity within the limits of its national jurisdiction; and

(b) In the case of processes and activities, regardless of where their effects occur, carried out under its jurisdiction or control, within the area of its national jurisdiction or beyond the limits of national jurisdiction.

Article 5. Cooperation

Each Contracting Party shall, as far as possible and as appropriate, cooperate with other Contracting Parties, directly or, where appropriate, through competent international organizations, in respect of areas beyond national jurisdiction and on other matters of mutual interest, for the conservation and sustainable use of biological diversity.

Article 6. General Measures for Conservation and Sustainable Use

Each Contracting Party shall, in accordance with its particular conditions and capabilities:

(a) Develop national strategies, plans or programmes for the conservation and sustainable use of biological diversity or adapt for this purpose existing strategies, plans or programmes which shall reflect, *inter alia*, the measures set out in this Convention relevant to the Contracting Party concerned; and

(b) Integrate, as far as possible and as appropriate, the conservation and sustainable use of biological diversity into relevant sectoral or cross-sectoral plans, programmes and policies.

Article 7. Identification and Monitoring

Each Contracting Party shall, as far as possible and as appropriate, in particular for the purposes of Articles 8 to 10:

(a) Identify components of biological diversity important for its conservation and sustainable use having regard to the indicative list of categories set down in Annex I;

(b) Monitor, through sampling and other techniques, the components of biological diversity identified pursuant to subparagraph (a) above, paying particular attention to those requiring urgent conservation measures and those which offer the greatest potential for sustainable use;

(c) Identify processes and categories of activities which have or are likely to have significant adverse impacts on the conservation and sustainable use of biological diversity, and monitor their effects through sampling and other techniques; and

(d) Maintain and organize, by any mechanism data, derived from identification and monitoring activities pursuant to subparagraphs (a), (b)and (c) above.

Article 8. *In-situ* Conservation

Each Contracting Party shall, as far as possible and as appropriate:

(a) Establish a system of protected areas or areas where special measures need to be taken to conserve biological diversity;

(b) Develop, where necessary, guidelines for the selection, establishment and management of protected areas or areas where special measures need to be taken to conserve biological diversity;

(c) Regulate or manage biological resources important for the conservation of biological diversity whether within or outside protected areas, with a view to ensuring their conservation and sustainable use;

(d) Promote the protection of ecosystems, natural habitats and the maintenance of viable populations of species in natural surroundings;

(e) Promote environmentally sound and sustainable development in areas adjacent to protected areas with a view to furthering protection of these areas;

(f) Rehabilitate and restore degraded ecosystems and promote the recovery of threatened species, *inter alia*, through the development and implementation of plans or other management strategies;

(g) Establish or maintain means to regulate, manage or control the risks associated with the use and release of living modified organisms resulting from biotechnology which are likely to have adverse environmental impacts that could affect the conservation and sustainable use of biological diversity, taking also into account the risks to human health;

(h) Prevent the introduction of, control or eradicate those alien species which threaten ecosystems, habitats or species;

(i) Endeavour to provide the conditions needed for compatibility between present uses and the conservation of biological diversity and the sustainable use of its components;

(j) Subject to its national legislation, respect, preserve and maintain knowledge, innovations and practices of indigenous and local communities embodying traditional lifestyles relevant for the conservation and sustainable use of biological diversity and promote their wider application with the approval and involvement of the holders of such knowledge, innovations and practices and encourage the equitable sharing of the benefits arising from the utilization of such knowledge, innovations and practices;

(k) Develop or maintain necessary legislation and/or other regulatory provisions for the protection of threatened species and populations;

(l) Where a significant adverse effect on biological diversity has been determined pursuant to Article 7, regulate or manage the relevant processes and categories of activities; and

(m) Cooperate in providing financial and other support for *in-situ* conservation outlined in subparagraphs (a) to (l) above, particularly to developing countries.

Article 9. *Ex-situ* Conservation

Each contracting Party shall, as far as possible and as appropriate and predominantly for the purpose of complementing *in-situ* measures:

(a) Adopt measures for the *ex-situ* conservation of components of biological diversity, preferably in the country of origin of such components;

(b) Establish and maintain facilities for *ex-situ* conservation of and research on plants, animals and micro-organisms, preferably in the country of origin of genetic resources;

(c) Adopt measures for the recovery and rehabilitation of threatened species and for their reintroduction into their natural habitats under appropriate conditions;

(d) Regulate and manage collection of biological resources from natural habitats for *ex-situ* conservation purposes so as not to threaten ecosystems and *in-situ* populations of species, except where special temporary *ex-situ* measures are required under subparagraph (c) above; and

(e) Cooperate in providing financial and other support for *ex-situ* conservation outlined in subparagraphs (a) to (d) above and in the establishment and maintenance of *ex-situ* conservation facilities in developing countries.

Article 10. Sustainable Use of Components of Biological Diversity

Each Contracting Party shall, as far as possible and as appropriate:

(a) Integrate consideration of the conservation and sustainable use of biological resources into national decision-making;

(b) Adopt measures relating to the use of biological resources to avoid or minimize adverse impacts on biological diversity;

(c) Protect and encourage customary use of biological resources in accordance with traditional cultural practices that are compatible with conservation or sustainable use requirements;

(d) Support local populations to develop and implement remedial action in degraded areas where biological diversity has been reduced; and

(e) Encourage cooperation between its governmental authorities and its private sector in developing methods for sustainable use of biological resources.

Article 11. Incentive Measures

Each Contracting Party shall, as far as possible and as appropriate, adopt economically and socially sound measures that act as incentives for the conservation and sustainable use of components of biological diversity.

Article 12. Research and Training

The Contracting Parties, taking into account the special needs of developing countries, shall:

(a) Establish and maintain programmes for scientific and technical education and training in measures for the identification, conservation and sustainable use of biological diversity and its components and provide support for such education and training for the specific needs of developing countries;

(b) Promote and encourage research which contributes to the conservation and sustainable use of biological diversity, particularly in developing countries, *inter alia*, in accordance with decisions of the Conference of the Parties taken in consequence of recommendations of the Subsidiary Body on Scientific, Technical and Technological Advice; and

(c) In keeping with the provisions of Articles 16, 18 and 20, promote and cooperate in the use of scientific advances in biological diversity research in developing methods for conservation and sustainable use of biological resources.

Article 13.   Public Education and Awareness

The Contracting Parties shall:

(a) Promote and encourage understanding of the importance of, and the measures required for, the conservation of biological diversity, as well as its propagation through media, and the inclusion of these topics in educational programmes; and

(b) Cooperate, as appropriate, with other States and international organizations in developing educational and public awareness programmes, with respect to conservation and sustainable use of biological diversity.

Article 14. Impact Assessment and Minimizing Adverse Impacts

Each Contracting Party, as far as possible and as appropriate, shall:

(a) Introduce appropriate procedures requiring environmental impact assessment of its proposed projects that are likely to have significant adverse effects on biological diversity with a view to avoiding or minimizing such effects and, where appropriate, allow for public participation in such procedures;

(b) Introduce appropriate arrangements to ensure that the environmental consequences of its programmes and policies that are likely to have significant adverse impacts on biological diversity are duly taken into account;

(c) Promote, on the basis of reciprocity, notification, exchange of information and consultation on activities under their jurisdiction or control which are likely to significantly affect adversely the biological diversity of other States or areas beyond the limits of national jurisdiction, by encouraging the conclusion of bilateral, regional or multilateral arrangements, as appropriate;

(d) In the case of imminent or grave danger or damage, originating under its jurisdiction or control, to biological diversity within the area under jurisdiction of other States or in areas beyond the limits of national jurisdiction, notify immediately the potentially affected States of such danger or damage, as well as initiate action to prevent or minimize such danger or damage; and

(e) Promote national arrangements for emergency responses to activities or events, whether caused naturally or otherwise, which present a grave and imminent danger to biological diversity and encourage international cooperation to supplement such national efforts and, where appropriate and agreed by the States or regional economic integration organizations concerned, to establish joint contingency plans.

2. The Conference of the Parties shall examine, on the basis of studies to be carried out, the issue of liability and redress, including restoration and compensation, for damage to biological diversity, except where such liability is a purely internal matter.

Article 15. Access to Genetic Resources

1. Recognizing the sovereign rights of States over their natural resources, the authority to determine access to genetic resources rests with the national governments and is subject to national legislation.

2. Each Contracting Party shall endeavour to create conditions to facilitate access to genetic resources for environmentally sound uses by other Contracting Parties and not to impose restrictions that run counter to the objectives of this Convention.

3. For the purpose of this convention the genetic resources being provided by a Contracting Party, as referred to in this Article and Articles 16 and 19, are only those that are provided by contracting Parties that are countries of origin of such resources or by the Parties that have acquired the genetic resources in accordance with this Convention.

4. Access, where granted, shall be on mutually agreed terms and subject to the provisions of this Article.

5. Access to genetic resources shall be subject to prior informed consent of the Contracting Party providing such resources, unless otherwise determined by that Party.

6. Each Contracting Party shall endeavour to develop and carry out scientific research based on genetic resources provided by other Contracting Parties with the full participation of, and where possible in, such Contracting Parties.

7. Each Contracting Party shall take legislative, administrative or policy measures, as appropriate, and in accordance with Articles 16 and 19 and, where necessary, through the financial mechanism established by Articles 20 and 21 with the aim of sharing in a fair and equitable way the results of research and development and the benefits arising from the commercial and other utilization of genetic resources with the Contracting Party providing such resources. Such sharing shall be upon mutually agreed terms .

Article 16. Access to and Transfer of Technology

1. Each Contracting Party, recognizing that technology includes biotechnology, and that both access to and transfer of technology among Contracting Parties are essential elements for the attainment of the objectives of this Convention, undertakes subject to the provisions of this Article to provide and/or facilitate access for and transfer to other Contracting Parties of technologies that are relevant to the conservation and sustainable use of biological diversity or make use of genetic resources and do not cause significant damage to the environment.

2. Access to and transfer of technology referred to in paragraph 1 above to developing countries shall be provided and/or facilitated under fair and most favourable terms, including on concessional and preferential terms where mutually agreed, and, where necessary, in accordance with the financial mechanism established by Articles 20 and 21. In the case of technology subject to patents and

other intellectual property rights, such access and transfer shall be provided on terms which recognize and are consistent with the adequate and effective protection of intellectual property rights. The application of this paragraph shall be consistent with paragraphs 3, 4 and 5 below.

3. Each Contracting Party shall take legislative, administrative or policy measures, as appropriate, with the aim that Contracting Parties, in particular those that are developing countries, which provide genetic resources are provided access to and transfer of technology which makes use of those resources, on mutually agreed terms, including technology protected by patents and other intellectual property rights, where necessary, through the provisions of Articles 20 and 21 and in accordance with international law and consistent with paragraphs 4 and 5 below.

4. Each Contracting Party shall take legislative, administrative or policy measures, as appropriate, with the aim that the private sector facilitates access to, joint development and transfer of technology referred to in paragraph 1 above for the benefit of both governmental institutions and the private sector of developing countries and in this regard shall abide by the obligations included in paragraphs 1, 2 and 3 above.

5. The Contracting Parties, recognizing that patents and other intellectual property rights may have an influence on the implementation of this Convention, shall cooperate in this regard subject to national legislation and international law in order to ensure that such rights are supportive of and do not run counter to its objectives.

Article 17. Exchange of Information

1. The Contracting Parties shall facilitate the exchange of information, from all publicly available sources, relevant to the conservation and sustainable use of biological diversity, taking into account the special needs of developing countries.

2. Such exchange of information shall include exchange of results of technical, scientific and socio-economic research, as well as information on training and surveying programmes, specialized knowledge, indigenous and traditional knowledge as such and in combination with the technologies referred to in Article 16, paragraph 1. It shall also, where feasible, include repatriation of information.

Article 18. Technical and Scientific Cooperation

1. The Contracting Parties shall promote international technical and scientific cooperation in the field of conservation and sustainable use of biological diversity, where necessary, through the appropriate international and national institutions.

2. Each Contracting Party shall promote technical and scientific cooperation with other Contracting Parties, in particular developing countries, in implementing this Convention, *inter alia*, through the development and implementation of national policies. In promoting such cooperation, special attention should be given

to the development and strengthening of national capabilities, by means of human resources development and institution building.

3. The Conference of the Parties, at its first meeting, shall determine how to establish a clearing-house mechanism to promote and facilitate technical and scientific cooperation.

4. The Contracting Parties shall, in accordance with national legislation and policies, encourage and develop methods of cooperation for the development and use of technologies, including indigenous and traditional technologies, in pursuance of the objectives of this Convention. For this purpose, the Contracting Parties shall also promote cooperation in the training of personnel and exchange of experts.

5. The Contracting Parties shall, subject to mutual agreement, promote the establishment of joint research programmes and joint ventures for the development of technologies relevant to the objectives of this Convention.

Article 19. Handling of Biotechnology and Distribution of Its Benefits

1. Each Contracting Party shall take legislative, administrative or policy measures, as appropriate, to provide for the effective participation in biotechnological research activities by those Contracting Parties, especially developing countries, which provide the genetic resources for such research, and where feasible in such Contracting Parties.

2. Each Contracting Party shall take all practicable measures to promote and advance priority access on a fair and equitable basis by Contracting Parties, especially developing countries, to the results and benefits arising from biotechnologies based upon genetic resources provided by those Contracting Parties. Such access shall be on mutually agreed terms.

3. The Parties shall consider the need for and modalities of a protocol setting out appropriate procedures, including, in particular, advance informed agreement, in the field of the safe transfer, handling and use of any living modified organism resulting from biotechnology that may have adverse effect on the conservation and sustainable use of biological diversity.

4. Each Contracting Party shall, directly or by requiring any natural or legal person under its jurisdiction providing the organisms referred to in paragraph 3 above, provide any available information about the use and safety regulations required by that Contracting Party in handling such organisms, as well as any available information on the potential adverse impact of the specific organisms concerned to the Contracting Party into which those organisms are to be introduced.

Article 20. Financial Resources

1. Each Contracting Party undertakes to provide, in accordance with its capabilities, financial support and incentives in respect of those national activities which are intended to achieve the objectives of this Convention, in accordance

with its national plans, priorities and programmes.

2. The developed country Parties shall provide new and additional financial resources to enable developing country Parties to meet the agreed full incremental costs to them of implementing measures which fulfil the obligations of this Convention and to benefit from its provisions and which costs are agreed between a developing country Party and the institutional structure referred to in Article 21, in accordance with policy, strategy, programme priorities and eligibility criteria and an indicative list of incremental costs established by the Conference of the Parties. Other Parties, including countries undergoing the process of transition to a market economy, may voluntarily assume the obligations of the developed country Parties. For the purpose of this Article, the Conference of the Parties, shall at its first meeting establish a list of developed country Parties and other Parties which voluntarily assume the obligations of the developed country Parties. The Conference of the Parties shall periodically review and if necessary amend the list. Contributions from other countries and sources on a voluntary basis would also be encouraged. The implementation of these commitments shall take into account the need for adequacy, predictability and timely flow of funds and the importance of burden-sharing among the contributing Parties included in the list.

3. The developed country Parties may also provide, and developing country Parties avail themselves of, financial resources related to the implementation of this Convention through bilateral, regional and other multilateral channels.

4. The extent to which developing country Parties will effectively implement their commitments under this convention will depend on the effective implementation by developed country Parties of their commitments under this convention related to financial resources and transfer of technology and will take fully into account the fact that economic and social development and eradication of poverty are the first and overriding priorities of the developing country Parties.

5. The Parties shall take full account of the specific needs and special situation of least developed countries in their actions with regard to funding and transfer of technology.

6. The Contracting Parties shall also take into consideration the special conditions resulting from the dependence on, distribution and location of, biological diversity within developing country Parties, in particular small island States.

7. Consideration shall also be given to the special situation of developing countries, including those that are most environmentally vulnerable, such as those with arid and semi-arid zones, coastal and mountainous areas.

Article 21. Financial Mechanism

1. There shall be a mechanism for the provision of financial resources to developing country Parties for purposes of this Convention on a grant or concessional basis the essential elements of which are described in this Article. The mechanism shall function under the authority and guidance of, and be accountable to, the Conference of the Parties for purposes of this Convention.

The operations of the mechanism shall be carried out by such institutional structure as may be decided upon by the Conference of the Parties at its first meeting. For purposes of this Convention, the Conference of the Parties shall determine the policy, strategy, programme priorities and eligibility criteria relating to the access to and utilization of such resources. The contributions shall be such as to take into account the need for predictability, adequacy and timely flow of funds referred to in Article 20 in accordance with the amount of resources needed to be decided periodically by the Conference of the Parties and the importance of burden-sharing among the contributing Parties included in the list referred to in Article 20, paragraph 2. Voluntary contributions may also be made by the developed country Parties and by other countries and sources. The mechanism shall operate within a democratic and transparent system of governance.

2. Pursuant to the objectives of this Convention, the Conference of the Parties shall at its first meeting determine the policy, strategy and programme priorities, as well as detailed criteria and guidelines for eligibility for access to and utilization of the financial resources including monitoring and evaluation on a regular basis of such utilization. The Conference of the Parties shall decide on the arrangements to give effect to paragraph 1 above after consultation with the institutional structure entrusted with the operation of the financial mechanism.

3. The Conference of the Parties shall review the effectiveness of the mechanism established under this Article, including the criteria and guidelines referred to in paragraph 2 above, not less than two years after the entry into force of this Convention and thereafter on a regular basis. Based on such review, it shall take appropriate action to improve the effectiveness of the mechanism if necessary.

4. The Contracting Parties shall consider strengthening existing financial institutions to provide financial resources for the conservation and sustainable use of biological diversity.

Article 22. Relationship with Other International Conventions

1. The provisions of this Convention shall not affect the rights and obligations of any Contracting Party deriving from any existing international agreement, except where the exercise of those rights and obligations would cause a serious damage or threat to biological diversity.

2. Contracting Parties shall implement this Convention with respect to the marine environment consistently with the rights and obligations of States under the law of the sea.

Article 23. Conference of the Parties

1. A Conference of the Parties is hereby established. The first meeting of the Conference of the Parties shall be convened by the Executive Director of the United Nations Environment Programme not later than one year after the entry into force of this convention. Thereafter, ordinary meetings of the Conference of the Parties shall be held at regular intervals to be determined by the Conference at its first meeting.

2. Extraordinary meetings of the Conference of the Parties shall be held at such other times as may be deemed necessary by the Conference, or at the written request of any Party, provided that, within six months of the request being communicated to them by the Secretariat, it is supported by at least one third of the Parties.

3. The Conference of the Parties shall by consensus agree upon and adopt rules of procedure for itself and for any subsidiary body it may establish, as well as financial rules governing the funding of the Secretariat. At each ordinary meeting, it shall adopt a budget for the financial period until the next ordinary meeting.

4. The Conference of the Parties shall keep under review the implementation of this Convention, and, for this purpose, shall:

(a) Establish the form and the intervals for transmitting the information to be submitted in accordance with Article 26 and consider such information as well as reports submitted by any subsidiary body;

(b) Review scientific, technical and technological advice on biological diversity provided in accordance with Article 25;

(c) Consider and adopt, as required, protocols in accordance with Article 28;

(d) Consider and adopt, as required, in accordance with Articles 29 and 30, amendments to this Convention and its annexes;

(e) Consider amendments to any protocol, as well as to any annexes thereto, and, if so decided, recommend their adoption to the parties to the protocol concerned;

(f) Consider and adopt, as required, in accordance with Article 30, additional annexes to this Convention;

(g) Establish such subsidiary bodies, particularly to provide scientific and technical advice, as are deemed necessary for the implementation of this convention;

(h) Contact, through the Secretariat, the executive bodies of conventions dealing with matters covered by this convention with a view to establishing appropriate forms of cooperation with them; and

(i) Consider and undertake any additional action that may be required for the achievement of the purposes of this Convention in the light of experience gained in its operation.

5. The United Nations, its specialized agencies and the International Atomic Energy Agency, as well as any State not Party to this Convention, may be represented as observers at meetings of the Conference of the Parties. Any other body or agency, whether governmental or nongovernmental, qualified in fields relating to conservation and sustainable use of biological diversity, which has informed the Secretariat of its wish to be represented as an observer at a meeting

of the Conference of the Parties, may be admitted unless at least one third of the Parties present object. The admission and participation of observers shall be subject to the rules of procedure adopted by the Conference of the Parties.

Article 24. Secretariat

A secretariat is hereby established. Its functions shall be:

(a) To arrange for and service meetings of the Conference of the Parties provided for in Article 23;

(b) To perform the functions assigned to it by any protocol;

(c) To prepare reports on the execution of its functions under this

Convention and present them to the Conference of the Parties;

(d) To coordinate with other relevant international bodies and, in particular to enter into such administrative and contractual arrangements as may be required for the effective discharge of its functions; and

(e) To perform such other functions as may be determined by the Conference of the Parties.

2. At its first ordinary meeting, the Conference of the Parties shall designate the secretariat from amongst those existing competent international organizations which have signified their willingness to carry out the secretariat functions under this Convention.

Article 25. Subsidiary Body on Scientific, Technical and Technological Advice

1. A subsidiary body for the provision of scientific, technical and technological advice is hereby established to provide the Conference of the Parties and, as appropriate, its other subsidiary bodies with timely advice relating to the implementation of this Convention. This body shall be open to participation by all Parties and shall be multi-disciplinary. It shall comprise government representatives competent in the relevant field of expertise. It shall report regularly to the Conference of the Parties on all aspects of its work.

2.   Under the authority of and in accordance with guidelines laid down by the Conference of the Parties, and upon its request, this body shall:

(a) Provide scientific and technical assessments of the status of biological diversity;

(b) Prepare scientific and technical assessments of the effects of types of measures taken in accordance with the provisions of this Convention;

(c) Identify innovative, efficient and state-of-the-art technologies and know-how relating to the conservation and sustainable use of biological diversity and advise on the ways and means of promoting development and/or transferring such technologies;

(d) Provide advice on scientific programmes and international cooperation in research and development related to conservation and sustainable use of biological diversity; and

(e) Respond to scientific, technical, technological and methodological questions that the Conference of the Parties and its subsidiary bodies may put to the body.

3. The functions, terms of reference, organization and operation of this body may be further elaborated by the Conference of the Parties.

Article 26. Reports

Each Contracting Party shall, at intervals to be determined by the Conference of the Parties, present to the Conference of the Parties, reports on measures which it has taken for the implementation of the provisions of this Convention and their effectiveness in meeting the objectives of this Convention.

Article 27. Settlement of Disputes

1. In the event of a dispute between contracting Parties concerning the interpretation or application of this Convention, the parties concerned shall seek solution by negotiation.

2. If the parties concerned cannot reach agreement by negotiation, they may jointly seek the good offices of, or request mediation by, a third party.

3. When ratifying, accepting, approving or acceding to this Convention, or at any time thereafter, a State or regional economic integration organization may declare in writing to the Depository that for a dispute not resolved in accordance with paragraph 1 or paragraph 2 above, it accepts one or both of the following means of dispute settlement as compulsory:

(a) Arbitration in accordance with the procedure laid down in Part 1 of Annex II;

(b) Submission of the dispute to the International Court of Justice.

4. If the parties to the dispute have not, in accordance with paragraph 3 above, accepted the same or any procedure, the dispute shall be submitted to conciliation in accordance with Part 2 of Annex II unless the parties otherwise agree.

5. The provisions of this Article shall apply with respect   to any protocol except as otherwise provided in the protocol concerned

Article 28. Adoption of Protocols

1. The Contracting Parties shall cooperate in the formulation and adoption of protocols to this Convention.

2. Protocols shall be adopted at a meeting of the Conference of the Parties.

3. The text of any proposed protocol shall be communicated to the Contracting

Parties by the Secretariat at least six months before such a meeting.

Article 29. Amendment of the Convention or Protocols

1. Amendments to this Convention may be proposed by any contracting Party. Amendments to any protocol may be proposed by any Party to that protocol.

2. Amendments to this Convention shall be adopted at a meeting of the Conference of the Parties. Amendments to any protocol shall be adopted at a meeting of the Parties to the Protocol in question. The text of any proposed amendment to this convention or to any protocol, except as may otherwise be provided in such protocol, shall be communicated to the Parties to the instrument in question by the secretariat at least six months before the meeting at which it is proposed for adoption. The secretariat shall also communicate proposed amendments to the signatories to this Convention for information.

3. The Parties shall make every effort to reach agreement on any proposed amendment to this Convention or to any protocol by consensus. If all efforts at consensus have been exhausted, and no agreement reached, the amendment shall as a last resort be adopted by a two-third majority vote of the Parties to the instrument in question present and voting at the meeting, and shall be submitted by the Depository to all Parties for ratification, acceptance or approval.

4. Ratification, acceptance or approval of amendments shall be notified to the Depository in writing. Amendments adopted in accordance with paragraph 3 above shall enter into force among Parties having accepted them on the ninetieth day after the deposit of instruments of ratification, acceptance or approval by at least two thirds of the Contracting Parties to this Convention or of the Parties to the protocol concerned, except as may otherwise be provided in such protocol. Thereafter the amendments shall enter into force for any other Party on the ninetieth day after that Party deposits its instrument of ratification, acceptance or approval of the amendments.

5. For the purposes of this Article, 'Parties present and voting' means Parties present and casting an affirmative or negative vote.

Article 30. Adoption and Amendment of Annexes

1. The annexes to this Convention or to any protocol shall form an integral part of the Convention or of such protocol, as the case may be, and, unless expressly provided otherwise, a reference to this Convention or its protocols constitutes at the same time a reference to any annexes thereto. Such annexes shall be restricted to procedural, scientific, technical and administrative matters.

2. Except as may be otherwise provided in any protocol with respect to its annexes, the following procedure shall apply to the proposal, adoption and entry into force of additional annexes to this convention or of annexes to any protocol:

(a) Annexes to this Convention or to any protocol shall be proposed and adopted according to the procedure laid down in Article 29;

(b) Any Party that is unable to approve an additional annex to this Convention or an annex to any protocol to which it is Party shall so notify the Depository, in writing, within one year from the date of the communication of the adoption by the Depository. The Depository shall without delay notify all Parties of any such notification received. A Party may at any time withdraw a previous declaration of objection and the annexes shall thereupon enter into force for that Party subject to subparagraph (c) below;

(c) On the expiry of one year from the date of the communication of the adoption by the Depository, the annex shall enter into force for all Parties to this Convention or to any protocol concerned which have not submitted a notification in accordance with the provisions of subparagraph (b) above.

3. The proposal, adoption and entry into force of amendments to annexes to this Convention or to any protocol shall be subject to the same procedure as for the proposal, adoption and entry into force of annexes to the Convention or annexes to any protocol.

4. If an additional annex or an amendment to an annex is related to an amendment to this Convention or to any protocol, the additional annex or amendment shall not enter into force until such time as the amendment to the Convention or to the protocol concerned enters into force.

Article 31. Right to Vote

1. Except as provided for in paragraph 2 below, each Contracting Party to any protocol shall have one vote.

2. Regional economic integration organizations, in matters within their competence, shall exercise their right to vote with a number of votes equal to the number of their member States which are Contracting Parties to this Convention or the relevant protocol. Such organizations shall not exercise their right to vote if their member States exercise theirs, and vice versa.

Article 32. Relationship between this Convention and Its Protocols

1. A State or a regional economic integration organization may not become a Party to a protocol unless it is, or becomes at the same time, a Contracting Party to this Convention.

2. Decisions under any protocol shall be taken only by the Parties to the protocol concerned. Any Contracting Party that has not ratified, accepted or approved a protocol may participate as an observer in any meeting of the parties to that protocol.

Article 33. Signature

This Convention shall be open for signature at Rio de Janeiro by all States and any regional economic integration organization from 5 June 1992 until 14 June 1992, and at the United Nations Headquarters in New York from 15 June

1992 to 4 June 1993.

Article 34. Ratification, Acceptance or Approval

1. This Convention and any protocol shall be subject to ratification, acceptance or approval by States and by regional economic integration organizations. Instruments of ratification, acceptance or approval shall be deposited with the Depository.

2. Any organization referred to in paragraph 1 above which becomes a Contracting Party to this Convention or any protocol without any of its member States being a Contracting Party shall be bound by all the obligations under the convention or the protocol, as the case may be. In the case of such organizations, one or more of whose member States is a Contracting Party to this Convention or relevant protocol, the organization and its member States shall decide on their respective responsibilities for the performance of their obligations under the convention or protocol, as the case may be. In such cases, the organization and the member States shall not be entitled to exercise rights under the convention or relevant protocol concurrently.

3. In their instruments of ratification, acceptance or approval, the organizations referred to in paragraph 1 above shall declare the extent of their competence with respect to the matters governed by the Convention or the relevant protocol. These organizations shall also inform the Depository of any relevant modification in the extent of their competence.

Article 35. Accession

1. This Convention and any protocol shall be open for accession by States and by regional economic integration organizations from the date on which the Convention or the protocol concerned is closed for signature. The instruments of accession shall be deposited with the Depository.

2. In their instruments of accession, the organizations referred to in paragraph 1 above shall declare the extent of their competence with respect to the matters governed by the Convention or the relevant protocol. These organizations shall also inform the Depository of any relevant modification in the extent of their competence.

3. The provisions of Article 34, paragraph 2, shall apply to regional economic integration organizations which accede to this Convention or any protocol.

Article 36. Entry Into Force

1. This Convention shall enter into force on the ninetieth day after the date of deposit of the thirtieth instrument of ratification, acceptance, approval or accession.

2. Any protocol shall enter into force on the ninetieth day after the date of deposit of the number of instruments of ratification, acceptance, approval or accession, specified in that protocol, has been deposited.

3. For each Contracting Party which ratifies, accepts or approves this Convention or accedes thereto after the deposit of the thirtieth instrument of ratification, acceptance, approval or accession, it shall enter into force on the ninetieth day after the date of deposit by such Contracting Party of its instrument of ratification, acceptance, approval or accession.

4. Any protocol, except as otherwise provided in such protocol, shall enter into force for a Contracting Party that ratifies, accepts or approves that protocol or accedes thereto after its entry into force pursuant to paragraph 2 above, on the ninetieth day after the date on which that Contracting Party deposits its instrument of ratification, acceptance, approval or accession, or on the date on which this Convention enters into force for that Contracting Party, whichever shall be the later.

5. For the purposes of paragraphs 1 and 2 above, any instrument deposited by a regional economic integration organization shall not be counted as additional to those deposited by member States of such organization.

Article 37. Reservations

No reservations may be made to this Convention.

Article 38. Withdrawals

1. At any time after two years from the date on which this Convention has entered into force for a Contracting Party, that Contracting Party may withdraw from the Convention by giving written notification to the Depository.

2. Any such withdrawal shall take place upon expiry of one y ear after the date of its receipt by the Depository, or on such later date as may be specified in the notification of the withdrawal.

3. Any Contracting Party which withdraws from this Convention shall be considered as also having withdrawn from any protocol to which it is party.

Article 39. Financial Interim Arrangements

Provided that it has been fully restructured in accordance with the requirements of Article 21, the Global Environment Facility of the United Nations Development Programme, the United Nations Environment Programme and the International Bank for Reconstruction and Development shall be the institutional structure referred to in Article 21 on an interim basis, for the period between the entry into force of this Convention and the first meeting of the Conference of the Parties or until the Conference of the Parties decides which institutional structure will be designated in accordance with Article 21.

Article 40. Secretariat Interim Arrangements

The secretariat to be provided by the Executive Director of the United Nations Environment Programme shall be the secretariat referred to in Article 24, paragraph 2, on an interim basis for the period between the entry into force of this convention and the first meeting of the Conference of the Parties.

Article 41. Depository

The Secretary-General of the United Nations shall assume the functions of Depository of this Convention and any protocols.

Article 42. Authentic Texts

The original of this Convention, of which the Arabic, Chinese, English, French, Russian and Spanish texts are equally authentic, shall be deposited with the Secretary-General of the United Nations.

IN WITNESS WHEREOF the undersigned, being duly authorized to that effect, have signed this Convention.

Done at Rio de Janeiro on this fifth day of June, one thousand nine hundred and ninety-two.

Annex I

IDENTIFICATION AND MONITORING

1. Ecosystems and habitats: containing high diversity, large numbers of endemic or threatened species, or wilderness; required by migratory species; of social, economic, cultural or scientific importance; or, which are representative, unique or associated with key evolutionary or other biological processes;

2. Species and communities which are: threatened; wild relatives of domesticated or cultivated species; of medicinal, agricultural or other economic value; or social, scientific or cultural importance; or importance for research into the conservation and sustainable use of biological diversity, such as indicator species; and

3. Described genomes and genes of social, scientific or economic importance.

Annex II

Part I

ARBITRATION

Article 1

The claimant party shall notify the secretariat that the parties are referring a dispute to arbitration pursuant to Article 27. The notification shall state the subject-matter of arbitration and include, in particular, the articles of the Convention or the protocol, the interpretation or application of which are at issue. If the parties do not agree on the subject matter of the dispute before the President of the tribunal is designated, the arbitral tribunal shall determine the subject matter. The secretariat shall forward the information thus received to all Contracting Parties to this Convention or to the protocol concerned.

Article 2

1. In disputes between two parties, the arbitral tribunal shall consist of three

members. Each of the parties to the dispute shall appoint an arbitrator and the two arbitrators so appointed shall designate by common agreement the third arbitrator who shall be the President of the tribunal. The latter shall not be a national of one of the parties to the dispute, nor have his or her usual place of residence in the territory of one of these parties, nor be employed by any of them, nor have dealt with the case in any other capacity.

2. In disputes between more than two parties, parties in the same interest shall appoint one arbitrator jointly by agreement.

3. Any vacancy shall be filled in the manner prescribed for the initial appointment.

Article 3

1. If the President of the arbitral tribunal has not been designated within two months of the appointment of the second arbitrator, the Secretary-General of the United Nations shall, at the request of a party, designate the President within a further two-month period.

2. If one of the parties to the dispute does not appoint an arbitrator within two months of receipt of the request, the other party may inform the Secretary-General who shall make the designation within a further two-month period.

Article 4

The arbitral tribunal shall render its decisions in accordance with the provisions of this Convention, any protocols concerned, and international law.

Article 5

Unless the parties to the dispute otherwise agree, the arbitral tribunal shall determine its own rules of procedure.

Article 6

The arbitral tribunal may, at the request of one of the parties, recommend essential interim measures of protection.

Article 7

The parties to the dispute shall facilitate the work of the arbitral tribunal and, in particular, using all means at their disposal, shall:

(a) Provide it with all relevant documents, information and facilities; and

(b) Enable it, when necessary, to call witnesses or experts and receive their evidence

Article 8

The parties and the arbitrators are under an obligation to protect the confidentiality of any information they receive in confidence during the proceedings of the arbitral tribunal.

Article 9

Unless the arbitral tribunal determines otherwise because of the particular circumstances of the case, the costs of the tribunal shall be borne by the parties to the dispute in equal shares. The tribunal shall keep a record of all its costs, and shall furnish a final statement thereof to the parties.

Article 10

Any Contracting Party that has an interest of a legal nature in the subject-matter of the dispute which may be affected by the decision in the case, may intervene in the proceedings with the consent of the tribunal.

Article 11

The tribunal may hear and determine counterclaims arising directly out of the subject-matter of the dispute.

Article 12

Decisions both on procedure and substance of the arbitral tribunal shall be taken by a majority vote of its members.

Article 13

If one of the parties to the dispute does not appear before the arbitral tribunal or fails to defend its case, the other party may request the tribunal to continue the proceedings and to make its award. Absence of a party or a failure of a party to defend its case shall not constitute a bar to the proceedings. Before rendering its final decision, the arbitral tribunal must satisfy itself that the claim is well founded in fact and law.

Article 14

The tribunal shall render its final decision within five months of the date on which it is fully constituted unless it finds it necessary to extend the time-limit for a period which should not exceed five more months.

Article 15

The final decision of the arbitral tribunal shall be confined to the subject-matter of the dispute and shall state the reasons on which it is based. It shall contain the names of the members who have participated and the date of the final decision. Any member of the tribunal may attach a separate or dissenting opinion to the final decision.

Article 16

The award shall be binding on the parties to the dispute. It shall be without appeal unless the parties to the dispute have agreed in advance to an appellate procedure.

Article 17

Any controversy which may arise between the parties to the dispute as regards the interpretation or manner of implementation of the final decision may be submitted by either party for decision to the arbitral tribunal which rendered it.

Part 2

CONCILIATION

Article 1

A conciliation commission shall be created upon the request of one of the parties to the dispute. The commission shall, unless the parties otherwise agree, be composed of five members, two appointed by each Party concerned and a President chosen jointly by those members.

Article 2

In disputes between more than two parties, parties in the same interest shall appoint their members of the commission jointly by agreement. Where two or more parties have separate interests or there is a disagreement as to whether they are of the same interest, they shall appoint their members separately.

Article 3

If any appointments by the parties are not made within two months of the date of the request to create a conciliation commission, the Secretary-General of the United Nations shall, if asked to do so by the party that made the request, make those appointments within a further two-month period.

Article 4

If a President of the conciliation commission has not been chosen within two months of the last of the members of the commission being appointed, the Secretary-General of the United Nations shall, if asked to do so by a party, designate a President within a further two-month period.

Article 5

The conciliation commission shall take its decisions by majority vote of its members. It shall, unless the parties to the dispute otherwise agree, determine its own procedure. It shall render a proposal for resolution of the dispute, which the parties shall consider in good faith.

Article 6

A disagreement as to whether the conciliation commission has competence shall be decided by the commission.

● ● ● ● ● ● ● ● ● ● ● ● ● ● ● ● ●